Communications in Computer and Information Science 794

Commenced Publication in 2007
Founding and Former Series Editors:
Phoebe Chen, Alfredo Cuzzocrea, Xiaoyong Du, Orhun Kara, Ting Liu,
Krishna M. Sivalingam, Dominik Ślęzak, Takashi Washio, Xiaokang Yang,
and Junsong Yuan

More information about this series at http://www.springer.com/series/7899

Vadim V. Strijov · Dmitry I. Ignatov ·
Konstantin V. Vorontsov (Eds.)

Intelligent Data Processing

11th International Conference, IDP 2016
Barcelona, Spain, October 10–14, 2016
Revised Selected Papers

 Springer

Editors
Vadim V. Strijov (ORCID)
Moscow Institute of Physics and Technology
Dolgoprudny, Russia

Konstantin V. Vorontsov (ORCID)
Yandex School of Data Analysis
Moscow, Russia

Dmitry I. Ignatov (ORCID)
National Research University Higher School
of Economics
Moscow, Russia

ISSN 1865-0929   ISSN 1865-0937 (electronic)
Communications in Computer and Information Science
ISBN 978-3-030-35399-5   ISBN 978-3-030-35400-8 (eBook)
https://doi.org/10.1007/978-3-030-35400-8

This Springer imprint is published by the registered company Springer Nature Switzerland AG
The registered company address is: Gewerbestrasse 11, 6330 Cham, Switzerland

Preface

The International Conference on Intelligent Information Processing: Theory and Applications is a premier forum for the data science researchers and professionals to discuss, distribute, and advance the state of research and development of the data analysis field. The conference facilitates the exchange of insights and innovations between the industry and academia, each represented by leaders in their respective fields. The IDP Conference has a rich history, starting from 1989 up till now. The conference offers research and industry tracks in the areas of Data Mining, Machine Learning, Big Data Analytics, Deep Learning, Computer Vision, Text Mining, Social Networks Analysis, and Data Science for Biology and Medicine.

The conference gathers plenty of highly recognized speakers from large industrial companies and academic research and educational organizations. The goal is to encourage ideas, success stories, and challenges sharing them among big data research and industry communities and establish connections between them and a talent pool of 200 experienced data scientists in our community. The invited speakers were selected from a wide list of recognizable candidates. Wonderful speakers came such as Paolo Rosso (Associate Professor at Universidad Politécnica de Valencia), Kamran Elahian (Chairman of Global Catalyst Partners), Konstantin V. Vorontsov (Professor at the Moscow Institute of Physics and Technology), Michael Levin (Chief Data Scientist at Yandex Data Factory), and Victor Lempitsky (Leader of Computer Vision Group at Skolkovo Institute of Science and Technology), among the other. Talks, fireside chats, research, and industry sessions as well as exhibitions were held during the event.

The Editorial Board and the conference Program Committee accepted papers representing the latest achievements in the theory and practice of intelligent data processing. Particular attention is given to the innovative solutions of the industry and business problems. Each paper contains a well-developed computational experiment including analysis and comparison aspects. The Editorial Board received 52 submissions and carefully selected 11 papers out them. The paper acceptance rate was 25%. Each submitted manuscript was examined whether it is prepared according to the guidelines and fits the scope of the conference. All the submitted manuscripts were subjected to the single-blind peer review process, participated at least by three independent reviewers, who remain anonymous throughout the process. The reviewers were matched to the papers according to their expertise. No conflicts of interest were allowed. The reviewers were also asked to evaluate whether the manuscript is original and of sufficient weight and interest.

This event would not have been possible without the participation of many people. We would like to thank everyone who contributed to the success of the conference. We would like especially acknowledge the dedicated contribution of members of the Organizing Committee. The conference was organized and supported by the Russian Foundation for Basic Research, Federal Research Center "Computer Science and Control" of the Russian Academy of Sciences, Moscow Institute of Physics and

Technology, Forecsys, Center of Forecasting Systems and Recognition, and Harbour. Space University Barcelona. We are grateful for the timely commitment of the Program Russian foundation for basic research and Reviewing Committee and all the external reviewers.

October 2016

Vadim V. Strijov
Dmitry Ignatov
Konstantin V. Vorontsov

Organization

Program Chairs

Konstantin Rudakov	Dorodnicyn Computing Centre of RAS, Moscow, Russia
Denis Zorin	New York University's Courant Institute of Mathematical Sciences, USA

Program Committee Secretary

Vadim V. Strijov	Moscow Institute of Physics and Technology, Russia

Program Committee

Mikhail Alexandrov	Autonomous University of Barcelona, Spain
Lorraine Baque	Autonomous University of Barcelona, Spain
Angels Catena	Autonomous University of Barcelona, Spain
Alexander Frey	Norment part University of Oslo, Finland
Ekaterrina Krymova	Universität Duisburg-Essen, Germany
Yuri Maximov	Los Alamos National Laboratory, USA
Konstantin Mertsalov	Harbour Space University Barcelona, Spain
Grigori Sidorov	National Polytechnic Institute of Mexico, Mexico
Roman Sologub	OBR Investments Limited, Cyprus
Iliya Tolstikhin	Max Planck Institute for Intelligent Systems, Germany
Liping Wang	Nanjing University of Aeronautics and Astronautics, China
Sergey Dvoenko	Tula State University, Russia
Edward Gimadi	Sobolev Institute of Mathematics of Siberian Branch of RAS (Novosibirsk) and Novosibirsk State University, Russia
Aleksandr Gornov	Institute of System Dynamics and Control Theory of Siberian Branch of RAS (Irkutsk), Russia
Olga Gromova	Ivanovo State Medical Academy, Russia
Aleksandr Kelmanov	Sobolev Institute of Mathematics of Siberian Branch of RAS (Novosibirsk), Russia
Mikhail Khachay	Yeltsin Ural Federal University and Institute of Mathematics and Mechanics of Ural Branch of RAS (Yekaterinburg), Russia
Leonid Mestetskiy	Lomonosov Moscow State University and Moscow Institute of Physics and Technology, Russia
Vadim Mottl	Dorodnicyn Computing Centre of RAS (Moscow) and Tula State University, Russia

Aleksandr Shananin	Moscow Institute of Physics and Technology, Russia
Viktor Soyfer	Samara State Aerospace University, Russia
Mikhail Ustinin	Institute of Mathematical Problems of Biology of RAS (Pushchino), Russia
Konstantin V. Vorontsov	Yandex School of Data Analysis and Dorodnicyn Computing Centre of RAS (Moscow), Russia

Reviewers

Aduenko Alexander
Bakhteev Oleg
Bunakova Vlada
Chicheva Marina
Chulichkov Alexey
Dudarenko Marina
Dvoenko Sergey
D'yakonov Alexander
Fedoryaka Dmitry
Frei Alexander
Gasnikov Alexander
Genrikhov Ivan
Gneushev Alexander
Goncharov Alexey
Graboviy Andriy
Ignat'ev Vladimir
Ignatov Dmitry
Inyakin Andrey
Isachenko Roman
Ishkina Shaura
Ivakhnenko Andrey
Karasikov Mikhail
Karkishchenko Alexander
Katrutsa Alexandr
Khachay Mikhail
Khritankov Anton
Kochetov Yury
Kopylov Andrey
Krasotkina Olga
Krymova Ekaterina
Kulunchakov Andrey
Kushnir Olesya
Kuzmin Arsenty
Kuznetsov Mikhail
Kuznetsova Rita
Lange Mikhail

Loukachevitch Natalia
Maksimov Yuri
Matrosov Mikhail
Matveev Ivan
Maysuradze Artchil
Mestetskiy Leonid
Mikheeva Anna
Mirkin Boris
Mnukhin Valeriy
Motrenko Anastasia
Murashov Dmirty
Nedel'ko Victor
Neychev Radoslav
Odinokikh Gleb
Panov Alexander
Panov Maxim
Potapenko Anna
Pushnyakov Alexey
Reyer Ivan
Riabenko Evgeniy
Sen'ko Oleg
Seredin Oleg
Stenina Maria
Strijov Vadim V.
Sulimova Valentina
Torshin Ivan
Trekin Anton
Turdakov Denis
Vetrov Dmitry
Vladimirova Maria
Vorontsov Konstantin V.
Yanina Anastasia
Zajtsev Alexey
Zharikov Ilya
Zhivotovskiy Nikita

Organising Committee Chairs

Yury Zhuravlev   Dorodnicyn Computing Centre of Russian Academy
                                              of Sciences, Moscow, Russia
Svetlana Velikanova   Harbour.Space University, Barcelona, Spain

Organising Committee Secretary

Yury Chehovich   Moscow Institute of Physics and Technology, Russia

Organising Committee

Tatyana Borisova   Center of Forecasting Systems and Recognition,
                                            Moscow, Russia
Andrey Gromov   Dorodnicyn Computing Centre of RAS, Moscow,
                                            Russia
Andrey Inyakin   Dorodnicyn Computing Centre of RAS, Moscow,
                                            Russia
Shaura Ishkina   Moscow Institute of Physics and Technology, Russia
Andrey Ivakhnenko   Antiplagiat, Moscow, Russia
Evgeniya Pomazkova   Moscow Institute of Physics and Technology, Russia
Lina Romashkova   Harbour.Space University, Barcelona, Spain
Aleksandr Tatarchuk   Forecsys, Moscow, Russia

Technical Editor

Shaura Ishkina   Moscow Institute of Physics and Technology, Russia

Sponsors

Russian Foundation for Basic Research
Moscow Institute of Physics and Technology (State University)
Forecsys
Center of Forecasting and Recognition Systems

Invited Talks and Tutorials

Robust Principal Component Analysis

Boris Polyak

Institute for Control Sciences Russian Academy of Sciences, Russia
boris ipu.ru

Abstract. The main trend of modern data analysis is to reduce huge databases to their low-dimensional approximations. Classical tool for this purpose is Principal Component Analysis (PCA). However, it is sensitive to outliers and other deviations from standard assumptions. There are numerous approaches to robust PCA. We propose two novel models. One is based on minimization of Huber-like distances from low-dimensional subspaces. A simple method for this nonconvex matrix optimization problem is proposed. The second is the robust version of maximum likelihood method for covariance and location estimation for contaminated multivariate Gaussian distribution; again we arrive at nonconvex vector-matrix optimization. Both methods are based on Reweighted Least Squares Approximations. They demonstrated fast convergence in simulations, however, statistical validation, as well as convergence behavior of both approaches, remain open problems.

Keywords: Low-dimensional approximations · Principal Component Analysis

Mining Intelligent Information in Twitter: Detecting Irony and Sarcasm

Paolo Rosso

Universitat Politècnica de València, Spain
prosso@dsic.upv.es

Abstract. There is a growing interest from the research community in investigating the impact of irony and sarcasm on sentiment analysis in social media. A sentiment analysis tool often processes information wrongly when dealing with ironic or sarcastic opinions. In fact, what is literally said is usually negated, moreover, in absence of an explicit negation marker. A task has been organised in 2015 at SemEval on sentiment analysis of figurative language in Twitter. In this talk I will describe how irony and sarcasm are employed in tweets and reviews in general and what are the recent state-of-the-art attempts for their automatic detection, for instance processing, explicit or implicit, affective information in tweets.

Keywords: Natural language processing · Sentiment analysis · Irony · Sarcasm

Big Data Platform and Analytics at Snapchat

Sina Sohangir

Snapchat, USA
sohangir@yahoo.com

Abstract. In this talk, I will go through our data platform at Snapchat which is based on Google Cloud. Then I present the type of problems we are trying to solve for ad targeting in general and for Snapchat in particular. I would look into Snapchat's social network and big data problems and solutions dealing with this specific social network.

Keywords: Snapchat · Ad targeting · Big data

Detecting Anomalies in the Real World

Alessandra Staglianò

The ASI, UK
alessandra.stagliano@unige.it

Abstract. Anomaly detection is a data science technique is used in many applications, from the IoT to finance. With the rise of the industrial Internet and the explosion of sensor data, businesses from transport to manufacturing are keen to develop predictive maintenance. Another key area where anomaly detection is important is identifying fraud in finance and within the social benefit system.

There are, however, a number of challenges when applying anomaly detection that are hindering progress. For a start, anomaly detection is a challenging problem by definition: defining and distinguishing between "normal" and an "anomaly" is often part of the problem statement. An anomaly is a relatively rare event and, hence, suffers from the accuracy paradox. Moreover, what is a good measure of success? Because of the nature of the problem, if the model misses all the anomalies, it will still be very accurate. The vastly different data types and preprocessing required, as well as the complex ensemble machine-learning methods needed, prove an additional challenge.

We illustrate these challenges and explain how to overcome them, with a practical application to credit card fraud detection.

Keywords: Anomaly detection · Fraud detection

Additively Regularized Topic Modeling for Searching Ethnical Discourse in Social Media

Konstantin V. Vorontsov

Yandex School of Data Analysis, Moscow Institute of Physics and Technology,
Lomonosov Moscow State University, Computing Centre of RAS, Russia
vokov@forecsys.ru

Abstract. Recently, social studies of the Internet have started to adopt various techniques of large-scale text mining for a variety of goals. One of such goals is the unsupervised discovery of topics related to ethnicity for early detection of ethnic conflicts emerging in social media. Probabilistic topic modeling used for such goals usually employs Bayesian inference for one of the numerous extensions of the Latent Dirichlet Allocation (LDA) model that has been widely popular over the last decade. However, recent research suggests that a non-Bayesian approach of additive regularization of topic models (ARTM) results in more control over the topics purity and interpretability, more flexibility for combining topic models, and faster inference.

In this work, we apply ARTM framework and BigARTM open-source software to a case study of mining ethnic content from the Russian-language blogosphere. We introduce a problem-specific combination of regularizers for ARTM and compare ARTM with LDA. The most important regularizer uses a vocabulary of a few hundred ethnonyms as seed words for ethnic-related topics. We conclude that ARTM is better suitable for mining rare topics, such as those on ethnicity, since it obviously finds larger numbers of relevant topics of higher or comparable quality.

Keywords: Topic modeling · Big ARTM · Ethnic discourse ·
Additive regularization

Image Synthesis with Deep Neural Networks

Victor Lempitsky

Skolkovo Institute of Science and Technology, Russia
lempitsky@skoltech.ru

Abstract. Using deep convolutional networks for pattern recognition in images has by now become a mature and well-known technology. More recently, there is a growing interest in using convolutional networks in a "reverse" mode, i.e. to synthesize images with certain properties rather than to recognize image content. In the talk, I will present several algorithmic results and application examples obtained for this very promising direction of research.

Keywords: Deep Learning · Convolutional networks · Image synthesis

Labelling Images Using Transfer Learning: An Application to Recommender Systems

Yannis Ghazouani

Dataiku, France

Abstract. Dataiku recently worked on an e-business vacation retailer recommender system. We created a meta model on top of classical recommender systems to optimise directly the probability of purchase. Along sales, clicks or descriptions of products data, we added image information in the recommender thanks to Deep Learning image recognition models. Since most companies cannot afford training neural networks, we followed a transfer learning approach.

The model we selected has been trained on the Places205 dataset and is based on the VGG16 architecture. We extracted last convolutional layer features from the mentioned VGG16 network and build a 2-layer TensorFlow model to learn the SUN397 labels from this intermediate features.

We then applied a Non-Negative Matrix Factorization (scikit-learn implementation) to reduce the dimension of the images representation space from around 600 categories to 30 variables and extracted some relevant topics from the images. Content-based image features were finally added to the meta model.

Keywords: Image labeling · Transfer learning · Feature extraction · Recommender systems

Contents

Machine Learning Theory with Applications

Overlapping Community Detection in Weighted Graphs: Matrix Factorization Approach

Konstantin Slavnov[1,2] and Maxim Panov[1(✉)]

[1] Skolkovo Institute of Science and Technology (Skoltech), Moscow, Russia
{k.slavnov,m.panov}@skoltech.ru
[2] Higher School of Economics, Moscow, Russia

Abstract. This work investigates the overlapping community detection problem. Recently, some efficient matrix factorization algorithms were proposed which can detect overlapping communities in unweighted graphs with millions of nodes. We expand these approaches to weighted graphs and develop a novel probabilistic model of overlapping community structure in weighted graphs. The resulting algorithm boils down to generalized matrix factorization with non-quadratic loss function. The comparison with the other methods shows that the proposed algorithm outperforms modern analogues.

Keywords: Overlapping community detection · Social graphs · Matrix factorization

1 Introduction

Communities are the basic structural units of the real world graphs [5]. They allow us to understand network structure and its properties. A big amount of work in computer science, statistics, physics and applied mathematics is devoted to detecting community structure in complex networks [4]. Community (group or cluster) is a closely linked vertex set sparsely connected with the rest of the graph vertices [5]. We can find clusters in social networks, in biochemical functional graphs [7], or in citation networks [2].

The problem of community detection has a variety of solution approaches in the case of non-overlapping communities and is quite well studied [4]. However, few methods exist for overlapping community detection, especially, in the case of weighted graphs. We note that many real world graphs are naturally weighted. Also, in many cases there exists some additional information, which can serve as edge weights. There are several methods such as MMSB [1], COPRA [6] and CFinder [11] which have been developed to solve the problem of overlapping community detection in weighted graphs. However, in general, this problem is not very well studied.

The common problem of overlapping community detection methods is poor scalability. The majority of methods can scale to networks with up to several

© Springer Nature Switzerland AG 2019
V. V. Strijov et al. (Eds.): IDP 2016, CCIS 794, pp. 3–14, 2019.
https://doi.org/10.1007/978-3-030-35400-8_1

thousands of nodes. Recently the BigCLAM method was proposed [14], which can scale to networks with tens of millions of edges. The method is based on the assumption that the more communities the vertices share, the greater is the probability of existing an edge between them. This assumption is supported by evidence from real data sets [14]. In BigCLAM the overlapping community detection problem is treated as generalized matrix factorization. It assumes that the adjacency matrix of a graph can be well approximated by a low-rank matrix. The most common approach to matrix factorization is to minimize the L_2 norm of the difference between adjacency matrix and its low-rank approximation. However, there exist criteria which better correspond to the properties of real world graphs than L_2 quality criterion. For example, the BigCLAM method utilizes the logistic type of criterion which works better for unweighted networks.

In this paper, we aim to create a scalable overlapping community detection method for weighted networks. We exploit the matrix factorization approach but propose the probabilistic model of the weighted graph with overlapping community structure. Similarly to BigCLAM, the model assumes that the pair of vertices sharing many communities has an edge between them with higher probability than the pair of vertices sharing no or few communities. The desired property is achieved using gamma distribution to generate edge weights. The resulting maximum likelihood problem is solved efficiently via block-coordinate descent method. The proposed algorithm shows the state-of-the-art performance on series of model and real problems. We conclude from the experiments that some community detection criteria can have only weak correspondence to the criteria based on the ground truth communities. This means that in the case when ground truth communities are not known, the analysis based on these criteria should be made very carefully to avoid misleading conclusions about the performance of the methods.

The remainder of the paper is organized as follows. In Sect. 2 we consider the unweighted graphs case and the BigCLAM method for the detection of overlapping communities. In Sect. 3 we turn to the case of weighted graphs and propose a new method for community detection based on the matrix factorization approach. Section 4 details the conducted experiments and reports the results. Section 5 concludes the study by discussing the main results and future research directions. The main notations used throughout the paper are summarized in Table 1.

2   Overlapping Community Detection in Graphs

Consider the case of community detection in unweighted and undirected graphs. Suppose that each vertex v of the graph G is affiliated to the community $c \in C$ with some non-negative weight F_{vc}. Zero value of F_{vc} means the absence of affiliation. This model can be represented by a biconnected graph as shown on Fig. 1.

Now we can define the edge occurrence probability

$$P(u, v) = 1 - \exp\left(-F_u F_v^{\mathrm{T}}\right), \tag{1}$$

Table 1. Main notations used in the paper.

Notation	Description
$G = (V, E)$	Graph as a set of vertices and edges
N	Number of vertices in a graph
K	Number of communities in a graph
$A \in \mathbb{R}_+^{N \times N}$	Adjacency matrix of a graph
$F \in \mathbb{R}_+^{N \times K}$	Matrix of communities affiliation weights
C	Set of all communities
$P(u, v)$	Probability of an edge (u, v)
$P((u, v) \mid c)$	Probability of an edge (u, v) under the condition that u and v belong to the community c
$l(F)$	Log-likelihood
$\mathcal{N}(u)$	1-neighborhood of vertex u
w_{uv}	Weight of an edge (u, v)

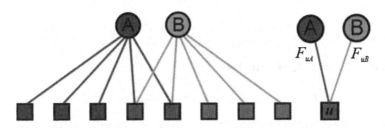

Fig. 1. Biconnected graph of BigClam model. The figure is taken from [14]. Community vertices are in the top row, the original graph vertices are in the bottom row. Edges represent community affiliations, while edges with zero weights are omitted for clarity.

where F_u is a row vector of affiliation weights $F_{uc} \geq 0$ of vertex u with all communities $c \in C$. The formula (1) is consistent with the general assumption: the more vertices the communities share, the higher the probability of existing an edge between them will be. Thus, we build the graph probability model with community structure, and we suppose that the observed graph was generated from it. We will discuss the probabilistic interpretation of this model in more details below, but first we introduce the BigClam approach and its underlying assumptions.

2.1 Cluster Affiliation Model for Big Networks (BigClam)

Consider the following assumptions on the random graph model.

1. Each vertex $v \in V$ belongs to a community $c \in C$ with some weight $F_{vc} \geq 0$.

2. If the vertices u and v are in the same community c, then the probability of an edge (u, v) is determined by

$$P\big((u, v) \mid c\big) = 1 - \exp(-F_{uc} \cdot F_{vc}).$$

3. Every community c generates an edge between u and v independently from other communities. The edge probability can be calculated by the formula for independent random variables:

$$P(u, v) = 1 - \prod_{c \in C}\big[1 - P\big((u, v) \mid c\big)\big] = 1 - \exp\big(-\sum_{c \in C} F_{uc} \cdot F_{vc}\big) = 1 - \exp\big(-F_u F_v^{\mathrm{T}}\big),$$

$$F = \{F_u\} = \{F_{uc}\} \in \mathbb{R}^{N \times K}.$$

The proposed model has a simple probabilistic interpretation. Suppose that there are latent random variables X_{uv}, which determine the existence of an edge between vertices of every pair of variables u and v. Edge is included in the graph only when $X_{uv} > 0$. Every community in the graph gives its independent contribution $X_{uv}^{(c)}$ to X_{uv}. Suppose that $X_{uv}^{(c)} \sim \mathrm{Pois}(F_{uc} \cdot F_{vc})$, where $F_{vc} \geq 0$ is community affiliation weight between vertex v and community c. Consequently

$$X_{uv} \sim \mathrm{Pois}\big(\sum_c F_{uc} \cdot F_{vc}\big) = \mathrm{Pois}\big(F_u F_v^{\mathrm{T}}\big).$$

The edge probability is equal to

$$P(u, v) = P(X_{uv} > 0) = 1 - \exp\big(-F_u F_v^{\mathrm{T}}\big), \tag{2}$$

which corresponds to the formula (1) given above. Note that the formula (2) fully determines the considered model.

The straightforward way to estimate matrix F is to use maximum likelihood approach. It is easy to derive that the log-likelihood $l(F)$ of the graph G with adjacency matrix A reads as

$$\begin{aligned} l(F) = l(F; \ A) &= \log \mathrm{P}(A \mid F) \\ &= \sum_{(u,v) \in E} \log(1 - \exp(-F_u F_v^{\mathrm{T}})) - \sum_{(u,v) \notin E} F_u F_v^{\mathrm{T}}. \end{aligned} \tag{3}$$

The probabilistic graph model described in this section was introduced in [14] under the name Cluster Affiliation Model for Big Networks (BigClam).

2.2 Optimization in the BigClam Model

We use block coordinate ascent algorithm to maximize $l(F)$ and get the matrix of community affiliation weights. Backtracking line search [3] is used to choose step size. One iteration of gradient projection method runs on each step.

One-step optimization is performed over F_u, while F_v for $u \neq v$ is fixed. The subproblem is convex and can be written as

$$\arg\max_{F_u \geq 0} l(F_u),$$

where

$$l(F_u) = \sum_{v \in \mathcal{N}(u)} \log\big(1 - \exp(-F_u F_v^{\mathrm{T}})\big) - \sum_{v \notin \mathcal{N}(u)} F_u F_v^{\mathrm{T}}$$

and $\mathcal{N}(u)$ is a set of vertex u neighbors. It is easy to verify that the gradient can be calculated as

$$\nabla l(F_u) = \sum_{v \in \mathcal{N}(u)} F_u \frac{\exp(-F_u F_v^{\mathrm{T}})}{1 - \exp(-F_u F_v^{\mathrm{T}})} - \sum_{v \notin \mathcal{N}(u)} F_v.$$

It would seem that the main computational burden (linear in number of vertices) is concentrated in the second term, were it not for the following decomposition:

$$\sum_{v \notin \mathcal{N}(u)} F_v = \sum_v F_v - F_u - \sum_{v \in \mathcal{N}(u)} F_v.$$

The $\sum_v F_v$ value is easy and cheap to update and maintain in memory. Hence, the computational complexity of one iteration becomes $O(\mathcal{N}(u))$, which is one of the main features of this method. The algorithm is scalable and can handle graphs with size up to 10^6 in a reasonable time.

After the optimization method has converged to some local optimum, we get the solution matrix F. To restore the community structure C based on it, we compare the matrix F with a threshold δ element-wise: if $F_{vc} > \delta$, then the vertex v is included into community c. We set δ equal to $\sqrt{-\log(1 - \varepsilon)}$ as suggested in [14]. For all our experiments we set $\varepsilon \sim 10^{-8}$.

2.3 Relation to the Matrix Factorization

Let us describe the relation of this problem to the matrix factorization. The generalized matrix factorization problem can be stated as the problem of finding a low-rank matrix $F \in \mathbb{R}_+^{N \times K}$, which is the best approximation of the adjacency matrix A in a sense of some discrepancy function D:

$$F = \arg\min_{F \geq 0} D\big(A, f(FF^{\mathrm{T}})\big).$$

Here and below the notation $F \geq 0$ means that matrix has all non-negative elements. Function $D(\cdot, f(\cdot))$ is measure of error between the matrix and its approximation. Such formulation allows us to consider the overlapping communities detection task as a non-negative matrix factorization problem with a general functional. In the BigClam case D is $-l(F; A)$ and $f(X) = 1 - \exp(-X)$. Function $f(\cdot)$ is called the link function and it transforms vertex affiliation weights to the edge occurrence probabilities. This functional is more suitable for binary matrices than the standard L_2 norm.

3   Weighted Graphs Models

Now we introduce a new model for the weighted case by defining a distribution over the latent variables X_{uv} and a weight generation rule. In the BigClam case the variables X_{uv} have Poisson distribution and, if they take a value greater than zero, an edge in the graph is created. This model can be extended to the case of integer weights by taking X_{uv} as a weight value. In the proposed model we select a continuous analogue of the Poisson distribution to make it able to produce a graph with continuous edge weights.

3.1   Gamma Model

We choose the gamma distribution $\Gamma(k, \theta)$ as its properties are well suited for the problem in question. Similarly to the previous section, we write the basic assumptions used in the model.

1. The probability of an edge to have a weight $w_{uv}^{(c)}$, under the condition that vertices belong to the same community c, is equal to

$$P\big(w_{uv}^{(c)} \mid c\big) \sim \Gamma\left(F_u F_v^{\mathrm{T}} + 1, 1\right).$$

2. Every community c generates an edge independently from each other. Then the edge weight in the graph is equal to

$$w_{uv} = \sum_{c \in C} w_{uv}^{(c)} \sim \Gamma\left(\sum_{c \in C} F_{uc} F_{vc} + 1, 1\right) = \Gamma\left(F_u F_v^{\mathrm{T}} + 1, 1\right).$$

As a result, we obtain the following log-likelihood:

$$l(F) = \sum_{(u,v)} \log P_\Gamma(w_{uv}) = \sum_{(u,v)} \left[-\log \Gamma\left(F_u F_v^{\mathrm{T}} + 1\right) + F_u F_v^{\mathrm{T}} \cdot \log w_{uv} - w_{uv}\right],$$

where the summation is taken over all nodes pairs in a graph. Optimization of this likelihood can be done by a simple gradient projection method. All we need is a new formula for the gradient, which can be written as

$$\nabla l(F_u) = -\sum_{v \in V} F_v \Psi\left(F_u F_v^{\mathrm{T}} + 1\right) - F_v \log w_{uv}, \tag{4}$$

where $\Psi(x) = \frac{\mathrm{d}}{\mathrm{d}x} \log(\Gamma(x))$ is the so-called digamma function. A small positive value κ was added to all the weights during the experiments to avoid zeros under the logarithm.

Unlike the original method, we can not apply the same reasoning with the complexity simplification of the gradient computation, because the sum in (4) is taken over all the edges in the graph. We have to recalculate the whole sum for each step and each F_u, so it takes time proportional to the total nodes number in the graph. The value of log-likelihood $l(F_u)$ also can be computed only in linear time, which is computationally expensive.

3.2   Sparse Gamma Model

The real world graphs are sparse and the gamma model cannot explain large number of missing edges in these graphs. Furthermore, the model is computationally expensive. Let us consider the following random graph generation model, which adds sparsity to the gamma model.

1. Matrix F and sparsity parameter $\gamma > 0$ are specified.
2. For any $u \in V$, $v \in V$: $v \neq u$ an edge (u, v) is added to the graph with probability $1 - \exp(-\gamma F_u F_v^{\mathrm{T}})$.
3. For every edge $(u, v) \in E$ the weight w_{uv} is generated from a corresponding gamma distribution: $w_{uv} \sim \Gamma\left(\sum_c F_{uc} F_{vc} + 1, 1\right)$.

The additional model parameter γ determines the sparsity of the graph. The smaller is its value, the sparser the adjacency matrix A will be.

Let us derive the likelihood formula for the described model. Denote the edge weight by w_{uv}, and binarized weights as a_{uv}, i.e. $a_{uv} = \mathbb{I}\left[w_{uv} > 0\right]$. Note, that the edge weight w_{uv} is non-zero if and only if $a_{uv} \neq 0$, hence

$$P(w_{uv} = 0 \mid a_{uv} = 0) = 1.$$

Taking into account these observations, the formula for log-likelihood can be derived, using the law of total probability:

$$
\begin{aligned}
l(F) = & \sum_{(u,v) \in E} \log P(w_{uv} \mid a_{uv} = 1) + \log P(a_{uv} = 1) \\
& + \sum_{(u,v) \notin E} \log P(w_{uv} = 0 \mid a_{uv} = 0) + \log P(a_{uv} = 0) \\
= & \sum_{(u,v) \in E} \log P_\Gamma(w_{uv}) + \sum_{(u,v) \in E} \log\left(1 - \exp(-\gamma F_u F_v^{\mathrm{T}})\right) - \gamma \sum_{(u,v) \notin E} F_u F_v^{\mathrm{T}}.
\end{aligned}
\tag{5}
$$

The first term is the log-likelihood of the previous model corresponding to edges with non-zero weights, and the last two terms represent the original BigClam model with $\sqrt{\gamma}F$ matrix. So, the resulting model is a combination of both. It retains all advantages of the BigClam model (see Sect. 2), including fast gradient calculation. It also takes into account the weighted edges and has an additional parameter γ, which binds the matrices of the gamma and the original models. In our experiments we did not explore the influence of the parameter γ and simply put $\gamma = 1$.

Since the model is a combination of two other models, just summing the gradients of the original models yields the new gradient. We note that the gradient of the gamma model includes the summation only over the neighbours of u. So, the original effective block-coordinate optimization scheme can be used.

3.3   Naive BigClam Generalization

In this section we introduce the simplest and the most intuitive generalization of the BigClam method, which will be called naive weighted BigClam. Indeed, the quality function (3) can be easily extended to weighted graphs by direct incorporation of weights. However, this approach has certain disadvantages in comparison to more advanced models from Sects. 3.1 and 3.2.

Denote edge weight (u, v) by w_{uv}. We change the BigClam quality function as follows:

$$l(F) = \sum_{(u,v) \in E} \log\left(1 - \exp\left(-\frac{F_u F_v^{\mathrm{T}}}{w_{uv}}\right)\right) - \sum_{(u,v) \notin E} F_u F_v^{\mathrm{T}}.$$

Only the first term has changed compared to the original model (3). Interaction weight $F_u F_v^{\mathrm{T}}$ between vertices u and v was normalized by the edge weight. It means, that the greater the weight w_{uv} is, the higher the value of weights F_u and F_v needs to be to explain it. Hence, the probability that vertices belong to the same community increases. Compared to the model (3), the probability of having an edge within the community c changes:

$$\mathrm{P}\big((u, v) \mid c, w_{uv}\big) = 1 - \exp\left(-\frac{F_{uc} F_{vc}}{w_{uv}}\right).$$

It is easy to notice a few drawbacks of this approach. First, the functional loses the symmetry of the original one: the edge presence or absence makes a different contribution to the functional. Second, it is hard to offer a probabilistic interpretation for the vertex weights like $X_{uv}^{(c)} \sim \mathrm{Pois}\left(\frac{F_{uc} \cdot F_{vc}}{w_{uv}}\right)$, since the distribution parameter is infinite in case of a missing edge. Moreover, we can not generate a graph from this model, because we need to know which edges are present and which are not, as well as expected weights of edges. Outlined difficulties in the interpretation give motivation to look for other ways to work with weighted data.

4   Experiments

4.1   Quality Functions

It is not trivial to measure community detection quality in the overlapping case, so we pay some attention to the ways of doing it. We use 2 quality measures: Modularity for Mixed Membership (*MixedModularity*) and Normalized Mutual Information (*NMI*). These two measures were originally developed for the non-overlapping communities case, so it is necessary to use their extensions. The generalization proposed in [13] is used for modularity, and the NMI extension can be found in [9].

We note that only the *NMI* uses ground truth information. So it is the most important quality function, because it directly measures the performance of clustering algorithm (while other measures focus on certain properties of the obtained clustering, which might not correlate with the measures based on ground truth information). However, we want to understand how other quality estimation methods correspond with the *NMI*.

4.2 Data

Most standard data sets are not suitable for testing the considered methods. Either there is no true overlapping community structure, or the graph is not weighted. Therefore, the main tests are conducted on a synthetic data set.

We followed the model introduced in [8] and used the code provided by the authors. The model has many parameters and we chose the basic set of parameter values as suggested in [10]. We conducted two experiments; see the corresponding parameter values in Table 2.

In the first experiment, we consider graph models with strong communities ($\mu_t = \mu_\omega = 0.1$) and weak communities ($\mu_t = \mu_\omega = 0.3$). We change the fraction of nodes belonging to multiple communities η from 0 to 0.5 and analyze the dependence of the quality on η to compare methods. When $\eta = 0$ communities do not overlap, and when $\eta = 0.5$ the half of vertices are found in the overlapping parts of the communities. For each set of parameter values, the results are averaged over 10 independent realizations of a graph.

In the second experiment, we look how the size of a graph influences the quality while keeping the number of communities in the graph approximately constant. The number of vertices in the graph is changed from 400 to 2000 and the average vertex degree is changed from 20 to 100 proportionally. We, again, average the results over 10 independent realizations of a graph.

Table 2. Parameter values in experiments.

	Experiment 1	Experiment 2	Description
N	1000	from 400 to 2000	Number of vertices
μ_t	0.1; 0.3	0.1	Mixing topology parameter
$k_{\max}$	50	$0.1N$	Maximum vertex degree
k	30	$0.05N$	Average vertex degree
μ_ω	0.1; 0.3	0.1	Mixing parameter for edge weights
η	from 0 to 0.5	0.1	Fraction of nodes belonging to multiple communities
ξ	2	2	Exponent parameter for weight distribution
τ_1	2	2	Minus exponent parameter for degree sequence
τ_2	2	2	Minus exponent parameter for community size distribution
o_m	2	2	Maximum number of community affiliations for a vertex

4.3 Results

In our experiments we consider the following methods.

1. *Sparse Gamma* is sparse gamma model described in Sect. 3.2.
2. *BigClam Weighted* is naive weighted BigClam described in Sect. 3.3.
3. *BigClam* is original BigClam [14] applied to the graph with binarized adjacency matrix.
4. *COPRA* is label propagation for overlapping community case [6].
5. *NMF* is a standard non-negative matrix factorization with quadratic norm.
6. *Walktrap* is method for non-overlapping community detection, based on random walks [12].
7. *Ground truth* is real community partition.

CFinder [11] was planned to be in the comparison list, but its speed is significantly lower than the speed of the methods above. It is not possible to obtain the results of its work within a reasonable time.

Figures 2 and 3 show the results of the first and second experiments respectively. The higher line means better method result in terms of the corresponding criterion. For the second experiment, we do not include the results for the *NMF* method, as they are significantly lower than results for the other methods.

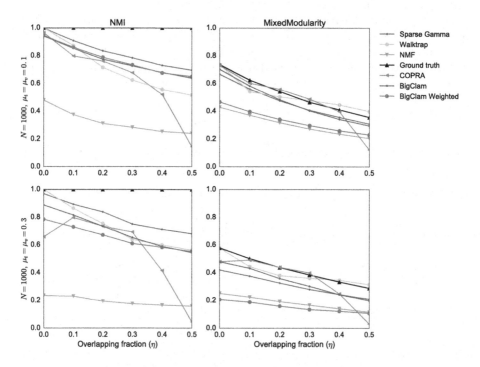

Fig. 2. Experiment results on synthetic data. The plots show the quality functions dependence on the parameter η. The rows correspond to 2 different parameter sets, other parameters were fixed as specified in Table 2. The columns correspond to different metrics: NMI and MixedModularity

For the first experiment the *NMI* graphs show that the *Sparse Gamma* model works better than other methods on these data. Only *Walktrap* outperforms *Sparse Gamma* model in case of disjoint communities, but it loses to the *Sparse Gamma* model as soon as the value of η deviates from zero.

In the second experiment, we can see that the most methods performance is not significantly affected by the size of a graph. Only *BigClam* shows noticeable decrease in quality. The *Sparse Gamma* model again outperforms other methods in terms of the *NMI*.

Note that in the terms of mixed modularity the results are not particularly illustrative. While ground truth communities have relatively high modularity values, the competing methods are ranked differently compared to the ranking based on the *NMI*. Thus, we can conclude that at least for the considered data mixed modularity is much less informative for comparing the obtained community structures than the *NMI*.

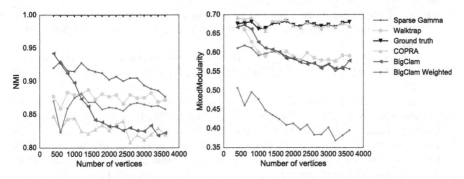

Fig. 3. Experiment results on synthetic data. The plots show the quality functions dependence on the number of vertices. The other parameters were fixed as specified in Table 2. The plots correspond to different metrics: NMI and MixedModularity

5   Discussion

In this paper we propose a new approach to the overlapping communities detection in weighted graphs based on sparse gamma model. The algorithm uses generalized non-negative matrix factorization approach, which was adapted to the case of weighted graphs. Importantly, the method combines good scalability with the state-of-the-art performance on several synthetic data sets, where the true community affiliations are known. Our experiments also show that if ground truth communities are not known, mixed modularity can be used for quality estimation. However, it allows to distinguish only between methods with significant gaps in performance, while the modularity values are not correlated with the NMI for closely competing methods. The future work is going to target additional experiments on the real data to make more grounded conclusions about the methods for overlapping community detection in weighted graphs.

Acknowledgments. The reported study was funded by RFBR according to the research project 18-37-00489.

References

1. Airoldi, E.M., Blei, D.M., Fienberg, S.E., Xing, E.P.: Mixed membership stochastic blockmodels. J. Mach. Learn. Res. **9**(Sep), 1981–2014 (2008). http://www.arxiv.org/abs/0709.2938
2. Backstrom, L., Huttenlocher, D., Kleinberg, J., Lan, X.: Group formation in large social networks: membership, growth, and evolution. In: Proceedings of the 12th ACM SIGKDD International Conference on Knowledge Discovery and Data Mining, pp. 44–54. ACM (2006). https://doi.org/10.1145/1150402.1150412
3. Boyd, S., Vandenberghe, L.: Convex Optimization. Cambridge University Press, Cambridge (2004). https://doi.org/10.1017/cbo9780511804441
4. Fortunato, S.: Community detection in graphs. Phys. Rep. **486**(3), 75–174 (2010). https://doi.org/10.1016/j.physrep.2009.11.002
5. Girvan, M., Newman, M.E.: Community structure in social and biological networks. Proc. Natl. Acad. Sci. **99**(12), 7821–7826 (2002). http://www.arxiv.org/pdf/cond-mat/0112110.pdf
6. Gregory, S.: Finding overlapping communities in networks by label propagation. New J. Phys. **12**(10), 103018 (2010). https://doi.org/10.1088/1367-2630/12/10/103018
7. Krogan, N.J., et al.: Global landscape of protein complexes in the yeast saccharomyces cerevisiae. Nature **440**(7084), 637–643 (2006). http://www.nature.com/nature/journal/v440/n7084/full/nature04670.html
8. Lancichinetti, A., Fortunato, S.: Benchmarks for testing community detection algorithms on directed and weighted graphs with overlapping communities. Phys. Rev. E **80**(1), 016118 (2009). https://doi.org/10.1103/physreve.80.016118
9. Lancichinetti, A., Fortunato, S., Kertész, J.: Detecting the overlapping and hierarchical community structure in complex networks. New J. Phys. **11**(3), 033015 (2009). https://arxiv.org/abs/0802.1218
10. Lu, Z., Sun, X., Wen, Y., Cao, G., La Porta, T.: Algorithms and applications for community detection in weighted networks. IEEE Trans. Parallel Distrib. Syst. **26**(11), 2916–2926 (2015). https://doi.org/10.1109/tpds.2014.2370031
11. Palla, G., Derényi, I., Farkas, I., Vicsek, T.: Uncovering the overlapping community structure of complex networks in nature and society. Nature **435**(7043), 814–818 (2005). https://doi.org/10.1038/nature03607
12. Pons, P., Latapy, M.: Computing communities in large networks using random walks. In: Yolum, I., Güngör, T., Gürgen, F., Özturan, C. (eds.) ISCIS 2005. LNCS, vol. 3733, pp. 284–293. Springer, Heidelberg (2005). https://doi.org/10.1007/11569596_31
13. Xie, J., Kelley, S., Szymanski, B.K.: Overlapping community detection in networks: the state-of-the-art and comparative study. ACM Comput. Surv. (CSUR) **45**(4), 43 (2013). https://doi.org/10.1145/2501654.2501657
14. Yang, J., Leskovec, J.: Overlapping community detection at scale: a nonnegative matrix factorization approach. In: Proceedings of the Sixth ACM International Conference on Web Search and Data Mining, pp. 587–596. ACM (2013). https://doi.org/10.1145/2433396.2433471

Reinforcement-Based Simultaneous Algorithm and Its Hyperparameters Selection

Valeria Efimova, Andrey Filchenkov$^{(\boxtimes)}$, and Anatoly Shalyto

ITMO University, St. Petersburg, Russia
efimova@rain.ifmo.ru, afilchenkov@corp.ifmo.ru, shalyto@mail.ifmo.ru

Abstract. There exist many algorithms for data analysis, especially for classification problems. To solve data analysis problem, a proper algorithm should be chosen, and also its hyperparameters should be selected. In this paper we present a new method for the simultaneous selection of an algorithm and its hyperparameters. In order to do so, we reduced this problem to the multi-armed bandit problem. We consider an algorithm as an arm and algorithm hyperparameters search during a fixed time as the corresponding arm play. We also suggest a problem-specific reward function. We performed the experiments on 10 real datasets and compare the suggested method with the existing one implemented in Auto-WEKA. The results show that our method is significantly better in most cases and never worse than the Auto-WEKA.

Keywords: Algorithm selection · Hyperparameter optimization · Multi-armed bandit · Reinforcement learning

1 Introduction

The goal of supervised learning is to find data model for a given dataset that allows to make the most accurate predictions. To build such a model, there exist lots of *learning algorithms*, especially in classification. These algorithms show various performances on different tasks. It impedes using a single universal algorithm to build data model for all existing datasets. The performance of most of these algorithms depends on *hyperparameters*, selection of which dramatically affects the algorithms performance.

Automated simultaneous selection of a learning algorithm and its hyperparameters is a sophisticated problem. Usually, this problem is divided into two subproblems that are solved independently: algorithm selection and hyperparameter optimization. The first is to select an algorithm from a set of algorithms (algorithm portfolio). The second is to find the best hyperparameters for preselected algorithm.

The first subproblem is typically solved by testing each of the algorithms with prechosen hyperparameters in the portfolio by many practitioners. Other

© Springer Nature Switzerland AG 2019
V. V. Strijov et al. (Eds.): IDP 2016, CCIS 794, pp. 15–27, 2019.
https://doi.org/10.1007/978-3-030-35400-8_2

methods are also in use, such as selecting algorithms randomly, by heuristics or using *k-fold cross-validation* [17]. But the last method requires running and then comparing all the algorithms. The other methods are not universally applicable. However, this subproblem has been in the scope of research interest for decades. Decision rules were used in several decades old papers on algorithm selection from a portfolio [2]. As an example, such rules are created to choose from 8 algorithms in [3].

Nowadays, more effective approaches exist, such as meta learning [1,10]. This approach is to reduce the algorithm selection problem to a supervised learning problem. It requires a training set of datasets D. For all $d \in D$, meta-feature vector is evaluated. Meta-features are useful characteristics of datasets, such as number of categorical or numerical features of an object $x \in d$, size of d and many others [8,9]. After that, all the algorithms are run on all the datasets $d \in D$. Thus, class labels are formed basing on empirical risk evaluation. Then, a meta-classifier is learnt on the prepared data with datasets as objects and best algorithms as labels. It is worth to note that it is better to solve this problem as a learning one to rank problem [7,20].

The second subproblem is hyperparameter optimization, that is to find hyperparameter vector for a learning algorithm that leads to the best performance of this algorithm for a given dataset. For example, hyperparameters of the Support Vector Machine (SVM) include kernel function and its hyperparameters; for a neural net, they include the number of hidden layers and the number of neurons in each of them. In practice, algorithms hyperparameters are usually chosen manually [13]. Moreover, sometimes the selection problem can be reduced to a simple optimization problem (primarily for statistical and regression algorithms), as, for instance, in [19]. However, this method is not universally applicable. Since hyperparameter optimization of classification algorithms is often applied manually, it requires a lot of time and does not lead to acceptable performance. There are several algorithms to solve the second subproblem automatically: Grid Search [4], Random Search [11], Stochastic Gradient Descent [6], Tree-structured Parzen estimator [5], and the Bayesian Optimization including Sequential Model-Based Optimization (SMBO) [18]. In [12], Sequential model-based algorithm configuration (SMAC) is introduced. It is based on SMBO algorithm. Another idea is implemented in predicting the best hyperparameter vector with meta-learning approach [16]. Reinforcement-based approach was used in [14] to operate several optimization threads with different settings.

Solution for an algorithm and its hyperparameters simultaneous selection is important for machine learning applications, but only a few papers are devoted to this search. Moreover, these papers consider only a special case.

One of the possible solutions is to build a huge set of algorithms with prechosen hyperparameters and select from it. This solution was implemented in [15], in which a set of about 300 algorithms with chosen hyperparameters was used. However, such pure algorithm selection approach cannot provide any insurance of these algorithms quality for a new problem. This set may simply not include a hyperparameter vector for one of the presented learning algorithms with the best performance.

Another possible solution is sequential optimization of hyperparameters for every learning algorithm in portfolio and selecting the best of them. This solution is implemented in the Auto-WEKA library [22], it allows to choose one of the 27 base learning algorithms, 10 meta-algorithms and 2 ensemble algorithms and optimize its hyperparameters with SMAC method simultaneously and automatically. This method is described in detail in [22]. It is clear, that if we use the method, then it takes enormous time and may be referred to as an exhaustive search (while, in fact, it is not due to the infinity of hyperparameter spaces).

The goal of this work is to suggest a method for simultaneous learning algorithm and its parameters selection, being faster than the exhaustive search without affecting found solution quality. In order to do so, we use multi-armed bandit-based approach.

The remainder of this paper is organized as follows. In Sect. 2, we describe in details the learning algorithm and its hyperparameter selection problem and its two subproblems. The suggested method, based on multi-armed bandit problem, is presented in Sect. 3. In Sect. 4, experiment results are presented and discussed. Section 5 concludes.

This paper extends a paper accepted to International Conference on Intelligent Data Processing: Theory and Applications 2016.

2   Problem Statement

Let Λ be a hyperparameter space related to a learning algorithm A. We will denote the algorithm with prechosen hyperparameter vector $\lambda \in \Lambda$ as A_λ.

Here is formal description of the algorithm selection problem. We are given a set of algorithms with chosen hyperparameters $\mathcal{A} = \{A^1_{\lambda_1}, \ldots A^m_{\lambda_m}\}$ and learning dataset $D = \{d_1, \ldots d_n\}$, where $d_i = (x_i, y_i)$ is a pair consisting of an object and its label. We should choose a parametrized algorithm $A^*_{\lambda_*}$ that is the most effective with respect to a quality measure Q. Algorithm efficiency is appraised by the use of dataset partition into learning and test sets with the further empirical risk estimation on the test set.

$$Q(A_\lambda, x) = \frac{1}{|D|} \sum_{x \in D} L(A_\lambda, x),$$

where $L(A_\lambda, x)$ is a loss function on object x, which is usually $L(A_\lambda, x) = [A_\lambda(x) \neq y(x)]$ for classification problems.

The algorithm selection problem thus is stated as the empirical risk minimization problem:

$$A^*_{\lambda_*} \in \underset{A^j_{\lambda_j} \in \mathcal{A}}{\operatorname{argmin}} Q(A^j_{\lambda_j}, D).$$

Hyperparameter optimization is the process of selecting hyperparameters $\lambda^* \in \Lambda$ of a learning algorithm A to optimize its performance. Therefore, we can write:

$$\lambda^* \in \underset{\lambda \in \Lambda}{\operatorname{argmin}}\, Q(A_\lambda, D).$$

In this paper, we consider the simultaneous algorithm selection and hyperparameters optimization. We are given learning algorithm set $\mathscr{A} = \{A^1, \ldots, A^k\}$. Each learning algorithm A^i is associated with hyperparameter space Λ^i. The goal is to find algorithm $A^*_{\lambda^*}$ minimizing the empirical risk:

$$A^*_{\lambda^*} \in \underset{A^j \in \mathscr{A}, \lambda \in \Lambda^j}{\operatorname{argmin}}\, Q(A^j_\lambda, D).$$

We assume that hyperparameter optimization is performed during the sequential hyperparameter optimization process. Let us give formal description. *Sequential hyperparameter optimization process* for a learning algorithm A^i:

$$\pi_i(t, A^i, \{\lambda^i_j\}^k_{j=0}) \to \lambda^i_{k+1} \in \Lambda^i.$$

It is a hyperparameter optimization method run on the learning algorithm A^i with time budget t, also it stores best found hyperparameter vectors within previous k iterations $\{\lambda_j\}^k_{j=0}$.

All of the hyperparameter optimization methods listed in the introduction can be described as a sequential hyperparameter optimization process, for instance, Grid Search or any of SMBO algorithm family including SMAC method, which is used in this paper.

Suppose that a sequential hyperparameter optimization process π_i is associated with each learning algorithm A_i. Then the previous problem can be solved by running all these processes. However, a new problem arises: time minimization problem for the best algorithm search. In practice, there is a similar problem that is more interesting in practical terms. It is the problem of finding the best algorithm by fixed time. Let us describe it formally.

Let T be a time budget for the best algorithm $A^*_{\lambda^*}$ searching. We should split T into intervals $T = t_1 + \cdots + t_m$ such that if we run process π_i with time budget t_i we will get minimal empirical risk.

$$\min_j Q(A^j_{\lambda_j}, D) \xrightarrow[(t_1, \ldots, t_m)]{} \min,$$

where $A^j \in \mathscr{A}, \lambda_j = \pi_j(t_j, A^j, \emptyset)$ and $t_1 + \ldots + t_m = T; t_i \geq 0 \forall i$.

3   Suggested Method

In this problem, the key source is hyperparameter optimization time limit T. Let us split it up to q equal small intervals t and call them *time budgets*. Now we can solve time budgets assignment problem. Let's have a look at our problem in a different way. For each time interval, we should choose a process to be run during this interval before this interval starts.

The quality that will be reached by an algorithm on a given dataset is a priori unknown. On the one hand, the time spent for searching hyperparameters of average learning algorithms is subtracted from the time spent to improve hyperparameters for the best learning algorithm. On the other hand, if the time will be spent only for tuning a single algorithm, we may miss better algorithms. Thus, since there is no marginal solution, the problem seems to be in finding a tradeoff between exploration (assigning time for tuning hyperparameters of different algorithms) and exploitation (assigning time for tuning hyperparameters of the current best algorithm). This tradeoff detection is the classical problem in reinforcement learning, a special case of which is multi-armed bandit problem [21]. We cannot assume that there is a hidden process for state transformation that affects performance of algorithms, thus we may assume that the environment is static.

Multi-armed bandit problem is a problem in which there are N bandit's arms. Playing each of the arms grants a certain reward. This reward is chosen according to unknown probability distribution, specific to this arm. At each iteration k, an agent chooses an arm a_i and get a reward $r(i, k)$. The agent's goal is to minimize the total loss by time T. In this paper, we use the following algorithms solving this problem [21]:

1. ε-greedy: on each iteration, average reward $\bar{r}_{a,t}$ is estimated for each arm a. Then the agent plays the arm with maximal average reward with probability $1 - \varepsilon$, and a random arm with probability ε. If you play each arm an infinite number of times, then the average reward converges to the real reward with probability 1.
2. UCB1: initially, the agent plays each arm once. On iteration t, it plays arm a_t that:
$$a_t \in \operatorname*{argmax}_{i=1..N} \overline{r_{i,t}} + \sqrt{\frac{2 \cdot \ln t}{n_i}},$$
where $\overline{r_{i,t}}$ is an average reward for arm i, n_i is the number of times arm i was played.
3. Softmax: initially, the agent plays each arm once. On iteration t, it plays arm a_i with probability:
$$p_{a_i} = \frac{e^{\bar{r}_i/\tau}}{\sum_{j=1}^{N} e^{r_j/\tau}},$$
where τ is positive temperature parameter.

In this paper, we associate arms with sequential hyperparameters optimization processes $\{\pi_i(t, A^i, \{\lambda_k\}_{k=0}^{q}) \rightarrow \lambda_{q+1}^i \in \Lambda^i\}_{i=0}^{m}$ for learning algorithms $\mathscr{A} = \{A^1, \ldots, A^m\}$. After playing arm $i = a_k$ at iteration k, we assign time budget t to a process π_{a_k} to optimize hyperparameters. When time budget runs out, we receive hyperparameter vector λ_k^i. Finally, when selected process stops, we evaluate the result using empirical risk estimated for process π_i at iteration k, that is $Q(A_{\lambda_k^i}^i, D)$.

Data: D is the given dataset q is the number of iterations,
t is time budget for one iteration,
$\{\pi_i\}_{i=1,\ldots,N}$ are sequential hyperparameter optimization processes.
Result: A_λ is algorithm with chosen hyperparameters

for $i = 1,\ldots,N$ do
$\quad\lambda_i \leftarrow \pi_i(t, A^i, \emptyset)$
$\quad e_i \leftarrow Q(A^i_{\lambda_i}, D)$
end
$best\_err \leftarrow \min_{i=1,\ldots,N} e_i$
$best\_proc \leftarrow \text{argmin}_{i=1,\ldots,N} e_i$
for $j = 1,\ldots,q$ do
$\quad i \leftarrow \text{MABSOLVER}(\{\pi_i\}_{i=1,\ldots,N})$
$\quad\lambda_i \leftarrow \pi_i(t, A^i, \{\lambda_k\}_{k=1}^j)$
$\quad e_i \leftarrow Q(A^i_{\lambda_i}, D)$
$\quad$if $e_i < best\_err$ then
$\quad\quad best\_err \leftarrow e_i$
$\quad\quad best\_proc \leftarrow i$
$\quad$end
end
return GETCONFIG(π_{best_proc})

Algorithm 1: MASSAH

The algorithm we name MASSAH (**M**ulti-**a**rmed **s**imultanous **s**election of **a**lgorithm and its **h**yperparameters) is presented listing 1. There, MABSOLVER is implementing a multi-armed bandit problem solution, GETCONFIG(i) is a function that returns $A^i_{\lambda_q}$, which is the best found configuration by q iterations to algorithm A^i.

The question we need to answer is how to define a reward function. The first (and simplest) way is to define a reward as the difference between current empirical risk and optimal empirical risk found during previous iterations. However, we meet several disadvantages. When the optimization process finds hyperparameters that lead to almost optimal algorithm performance, the reward will be extremely small. Also, the selection of such a reward function does not seem to be a good option for MABs, since probability distribution will depend on the number of iterations.

In order to find such a reward function, that the corresponding probability distribution will not change during the algorithm performance, we apply a little trick. Instead of defining reward function itself, we will define an average reward function. In order to do so, we use SMAC algorithm features.

Let us describe SMAC algorithm. At each iteration, a set of current optimal hyperparameter vectors is known for each algorithm. A local search is applied to find hyperparameter vectors which have distinction in one position with an optimal vector and improve algorithm quality. These hyperparameter vectors are added to the set. Moreover, some random hyperparameter vectors are added to the set. Then selected configurations (the algorithms with their hyperparameters) are sorted by *expected improvement* (EI). Some of the best configurations are run after that.

As in SMAC, we use empirical risk expectation at iteration k: $E_t(Q(A^i_{\lambda^i_k}, D))$, where $Q(A^i_{\lambda^i_k}, D)$ is empirical risk value reached by process π_i on dataset D at iteration k.

Table 1. Datasets description

Dataset	Number of categorical features	Number of numerical features	Number of classes	Number of objects in training set	Number of objects in test set
Dexter	0	20000	2	420	180
German Credit	13	7	2	700	300
Dorothea	0	100000	2	805	345
Yeast	0	8	10	1039	445
Secom	0	590	2	1097	470
Semeion	0	256	10	1116	477
Car	6	0	4	1210	518
KR-vs-KP	36	0	2	2238	958
Waveform	0	40	3	3500	1500
Shuttle	38	192	2	35000	15000

Note that process π_i optimizes hyperparameters for empirical risk minimization, but a multi-armed bandit problem is a maximization problem. Therefore, we define an average reward function as:

$$\bar{r}_{i,(k)} = \frac{Q_{max} - E_{(k)}(Q(A^i_{\lambda^i_k}, D))}{Q_{max}},$$

where Q_{max} is the maximal empirical risk that was achieved on a given dataset.

4   Experiments

Since Auto-WEKA implements the only existing solution, we choose it for comparison. Experiments were performed on 10 different real datasets with a predefined split into training and test data from UCI repository[1]. These datasets characteristics are presented in Table 1.

The suggested approach allows to use any hyperparameter optimization method. In order to perform comparison properly, we use SMAC method that is used by Auto-WEKA. We consider 6 well-known classification algorithms:

---

[1] http://www.cs.ubc.ca/labs/beta/Projects/autoweka/datasets/.

k Nearest Neighbors (4 categorical and 1 numerical hyperparameters), Support Vector Machine (4 and 6), Logistic Regression (0 and 1), Random Forest (2 and 3), Perceptron (5 and 2), and C4.5 Decision Tree (6 and 2).

As we have previously stated, we are given time T to find the solution of the main problem. The suggested method requires splitting T into small equal intervals t. We give the small interval to a selected process π_i at each iteration. We compare the method performance for different time budget t values to find the optimal value. We consider time budgets from 10 to 60 s with 3 second step. After that we run the suggested method on 3 datasets Car, German Credits, KRvsKP described above. We use 4 solutions of the multi-armed bandit problem: UCB1, 0.4-greedy, 0.6-greedy, Softmax. We run each configuration 3 times. The results show no regularity, so we assume time budget t as 30 s.

In the quality comparison, we consider suggested method with the different multi-armed bandit problem solutions: UCB1, 0.4-greedy, 0.6-greedy, Softmax with the naïve reward function, and two solutions $UCB1_{E(Q)}$, $Softmax_{E(Q)}$ with the suggested reward function. Time budget on iteration is $t = 30$ s, the general time limitation is $T = 3$ h $= 10800$ s. We run each configuration 12 times with random seeds of SMAC algorithm. Auto-WEKA is also limited to 3 h and selects one of the algorithms we specified above. The experiment results are shown in Table 2.

Table 2. Comparison of Auto-WEKA and suggested methods for selecting classification algorithm and its hyperparameters for the given dataset. We performed 12 independent runs of each configuration and report the smallest empirical risk Q achieved by Auto-WEKA and the suggested method variations. We highlight with bold entries that are minimal for the given dataset

Dataset	AutoWEKA	UCB1	0.4-greedy	0.6-greedy	Softmax	$UCB1_{E(Q)}$	$Softmax_{E(Q)}$
Car	0.3305	**0.1836**	**0.1836**	**0.1836**	**0.1836**	**0.1836**	**0.1836**
Yeast	34.13	**29.81**	**29.81**	33.65	**29.81**	**29.81**	**29.81**
KR-vs-KP	0.2976	**0.1488**	**0.1488**	**0.1488**	**0.1488**	**0.1488**	**0.1488**
Semeion	4.646	**1.786**	**1.786**	**1.786**	**1.786**	**1.786**	**1.786**
Shuttle	**0.00766**	0.0115	0.0115	**0.00766**	0.0115	**0.0076**	**0.0076**
Dexter	7.143	2.38	2.381	2.381	2.381	2.381	**0.16**
Waveform	11.28	**8.286**	**8.286**	**8.286**	**8.286**	**8.286**	**8.286**
Secom	4.545	**3.636**	4.545	4.545	**3.636**	**3.636**	**3.636**
Dorothea	6.676	4.938	4.958	4.938	4.938	4.32	**2.469**
German Credits	19.29	**14.29**	**14.29**	15.71	**14.29**	**14.29**	**14.29**

The results show that the suggested method is significantly better in most cases than Auto-WEKA on all the 10 datasets, because its variations reach the smallest empirical risk. There is no fundamental difference between the results of the suggested method variations. Nevertheless, algorithms $UCB1_{E(Q)}$ and $Softmax_{E(Q)}$, which use the suggested reward function, achieved the smallest empirical risk in most cases.

The experiment results show that the suggested approach improves the existing solution of the simultaneous learning algorithm and its hyperparameters selection problem. Moreover, the suggested approach does not impose restrictions on a hyperparameter optimization process, so the search is performed on the entire hyperparameters space for each learning algorithm. It is significant, that the suggested method allows to select a learning algorithm with hyperparameters, whose quality is not worse than Auto-WEKA outcome quality.

An illustration of algorithm performance comparison is presented in Fig. 1. As we can see, some algorithms had initially good shots at the beginning, while other have been improving it performance over time. However, being stuck in suboptimal solutions is an often situation, as it can be seem from Fig. 2.

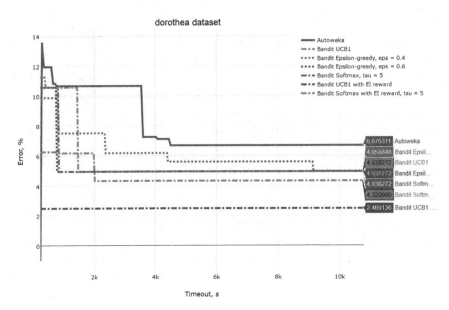

Fig. 1. Dependency of error on performance time for the best of 12 runs on Dorothea dataset

Plots depicted in Figs. 3 and 4 show dynamics of the best found solution time. In the first case the algorithm spent some time to find the best model, namely Perceptron, but shortly after it was not able to improve its performance. In contrast, in the second case the algorithm quickly found the best model namely SVM, but spent a lot of time for finding its best configuration.

We claim that the suggested method is statistically not worse than Auto-WEKA. To prove this, we carried out Wilcoxon signed-rank test. In experiments we use 10 datasets which leads to an appropriate number of pairs. Moreover, other Wilcoxon test assumptions are carried. Therefore, we have 6 test checks: comparison of Auto-WEKA and each variation of the suggested method. Since the number of samples is 10, we have meaningful results when untypical results sum $T < T_{0,01} = 5$. We consider a minimization problem, so we test only the best of 12 runs for each dataset. Finally, we have $T = 3$ for the ε-greedy algorithms and $T = 1$ for the others. This proves the statistical significance of the obtained results.

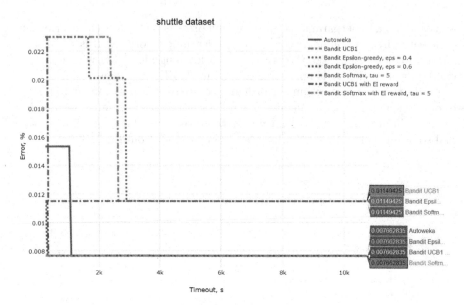

Fig. 2. Error dependency on performance time for the best of 12 runs on Shuttle dataset

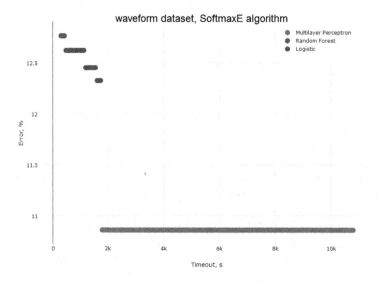

Fig. 3. The best model performance in dependence on performance time on Dorothea dataset

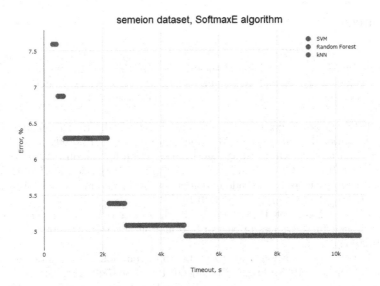

Fig. 4. The best model performance in dependence on performance time on Shuttle dataset

5   Conclusions

In this paper, we suggest and examine a new solution to the actual problem of an algorithm and its hyperparameters simultaneous selection. The proposed approach is based on a multi-armed bandit problem solution. We suggest a new reward function exploiting hyperparameter optimization method properties. The suggested function is better than the naïve function in applying a multi-armed bandit problem solutions to solve the main problem. The experiment result shows that the suggested method outperforms the existing method implemented in Auto-WEKA.

The suggested method can be improved by applying meta-learning in order to evaluate algorithm quality to preprocess a given dataset before running any algorithm. This evaluation can be used as a prior knowledge of an algorithm reward. Moreover, we can add a context vector to hyperparameters optimization process and use solutions of a contextual multi-armed bandit problem. We can select some datasets by meta-learning and then get the empirical risk estimate and use it as context.

Authors would like to thank Vadim Strijov and unknown reviewers of IDP'16 conference for useful comments. The research was supported by the Government of the Russian Federation (grant 074-U01) and the Russian Foundation for Basic Research (project no. 16-37-60115).

References

1. Abdulrahman, S.M., Brazdil, P., van Rijn, J.N., Vanschoren, J.: Algorithm selection via meta-learning and sample-based active testing. In: European Conference on Machine Learning and Principles and Practice of Knowledge Discovery in Databases; International Workshop on Meta-Learning and Algorithm Selection. University of Porto (2015)
2. Aha, D.W.: Generalizing from case studies: a case study. In: Proceedings of the 9th International Conference on Machine Learning, pp. 1–10 (1992)
3. Ali, S., Smith, K.A.: On learning algorithm selection for classification. Appl. Soft Comput. **6**(2), 119–138 (2006). https://doi.org/10.1016/j.asoc.2004.12.002
4. Bergstra, J., Bengio, Y.: Random search for hyper-parameter optimization. J. Mach. Learn. Res. **13**(1), 281–305 (2012)
5. Bergstra, J.S., Bardenet, R., Bengio, Y., Kégl, B.: Algorithms for hyper-parameter optimization. In: Advances in Neural Information Processing Systems, pp. 2546–2554 (2011)
6. Bottou, L.: Online learning and stochastic approximations. On-line Learn. Neural Netw. **17**(9), 142 (1998). https://doi.org/10.1017/cbo9780511569920.003
7. Brazdil, P.B., Soares, C., Da Costa, J.P.: Ranking learning algorithms: using IBL and meta-learning on accuracy and time results. Mach. Learn. **50**(3), 251–277 (2003)
8. Castiello, C., Castellano, G., Fanelli, A.M.: Meta-data: characterization of input features for meta-learning. In: Torra, V., Narukawa, Y., Miyamoto, S. (eds.) MDAI 2005. LNCS (LNAI), vol. 3558, pp. 457–468. Springer, Heidelberg (2005). https://doi.org/10.1007/11526018_45
9. Filchenkov, A., Pendryak, A.: Datasets meta-feature description for recommending feature selection algorithm. In: Artificial Intelligence and Natural Language and Information Extraction, Social Media and Web Search FRUCT Conference (AINL-ISMW FRUCT), pp. 11–18. IEEE (2015). https://doi.org/10.1109/ainl-ismw-fruct.2015.7382962
10. Giraud-Carrier, C., Vilalta, R., Brazdil, P.: Introduction to the special issue on meta-learning. Mach. Learn. **54**(3), 187–193 (2004). https://doi.org/10.1023/b:mach.0000015878.60765.42
11. Hastie, T., Tibshirani, R., Friedman, J., Franklin, J.: The elements of statistical learning: data mining, inference and rediction. Math. Intell. **27**(2), 83–85 (2005). https://doi.org/10.1007/bf02985802
12. Hutter, F., Hoos, H.H., Leyton-Brown, K.: Sequential model-based optimization for general algorithm configuration. In: Coello, C.A.C. (ed.) LION 2011. LNCS, vol. 6683, pp. 507–523. Springer, Heidelberg (2011). https://doi.org/10.1007/978-3-642-25566-3_40
13. Hutter, F., Lücke, J., Schmidt-Thieme, L.: Beyond manual tuning of hyperparameters. KI-Künstliche Intell. **29**(4), 329–337 (2015). https://doi.org/10.1007/s13218-015-0381-0
14. Jamieson, K., Talwalkar, A.: Non-stochastic best arm identification and hyperparameter optimization. JMLR **41**, 240–248 (2015)
15. Leite, R., Brazdil, P., Vanschoren, J.: Selecting classification algorithms with active testing. In: Perner, P. (ed.) MLDM 2012. LNCS (LNAI), vol. 7376, pp. 117–131. Springer, Heidelberg (2012). https://doi.org/10.1007/978-3-642-31537-4_10

16. Mantovani, R.G., Rossi, A.L., Vanschoren, J., Carvalho, A.C.P.D.L., et al.: Meta-learning recommendation of default hyper-parameter values for SVMs in classifications tasks. In: European Conference on Machine Learning and Principles and Practice of Knowledge Discovery in Databases; International Workshop on Meta-Learning and Algorithm Selection. University of Porto (2015)
17. Rodriguez, J.D., Perez, A., Lozano, J.A.: Sensitivity analysis of k-fold cross validation in prediction error estimation. IEEE Trans. Pattern Anal. Mach. Intell. **32**(3), 569–575 (2010). https://doi.org/10.1109/tpami.2009.187
18. Snoek, J., Larochelle, H., Adams, R.P.: Practical Bayesian optimization of machine learning algorithms. In: Advances in Neural Information Processing Systems, pp. 2951–2959 (2012)
19. Strijov, V., Weber, G.W.: Nonlinear regression model generation using hyperparameter optimization. Comput. Math. Appl. **60**(4), 981–988 (2010). https://doi.org/10.1016/j.camwa.2010.03.021
20. Sun, Q., Pfahringer, B.: Pairwise meta-rules for better meta-learning-based algorithm ranking. Mach. Learn. **93**(1), 141–161 (2013). https://doi.org/10.1007/s10994-013-5387-y
21. Sutton, R.S., Barto, A.G.: Reinforcement Learning: An Introduction. MIT Press, Cambridge (1998)
22. Thornton, C., Hutter, F., Hoos, H.H., Leyton-Brown, K.: Auto-WEKA: automated selection and hyper-parameter optimization of classification algorithms. CoRR, abs/1208.3719 (2012). https://doi.org/10.1145/2487575.2487629

An Information Approach to Accuracy Comparison for Classification Schemes in an Ensemble of Data Sources

Mikhail Lange$^{(\boxtimes)}$, Sergey Ganebnykh, and Andrey Lange

Federal Research Center "Computer Science and Control"
of Russian Academy of Sciences, 44-2 Vavilov Str, Moscow, Russia
lange_mm@ccas.ru

Abstract. An accuracy of multiclass classifying the collections of objects taken from a given ensemble of data sources is investigated using the average mutual information between the datasets of the sources and a set of the classes. We consider two fusion schemes, namely WMV (Weighted Majority Vote) scheme based on a composition of decisions on the objects of the individual sources and GDM (General Dissimilarity Measure) scheme which uses a composition of metrics in datasets of the sources. For a given metric classification model, it is proved that the weighted mean value of the average mutual information per one source in WMV scheme is smaller to the similar mean in GDM scheme. Using a lower bound to the appropriate rate distortion function, it is shown that the lower bounded error probability in WMV scheme exceeds the similar error probability in GDM scheme. This theoretical result is confirmed by a computing experiment on face recognition of HSI color images giving the ensemble of H, S, and I sources.

Keywords: Multiclass classification · Ensemble of sources · Fusion scheme · Composition of decisions · Composition of metrics · Average mutual information · Error probability

1 Introduction

The data fusion problem appears in many multiclass classification tasks, in which the decisions are made on the submitted collections of the objects taken from the sources of different modalities. The ensemble of the sources produces the composite objects as the collections of different modality objects belonging to the same class. An example is the ensemble of biometric images such as faces, fingerprints, signatures, palms, irises and the like. In this ensemble, the composite objects are given by the collections of the same person images taken with one from each source. For the composite objects, a classification error probability decreases with increasing a number of the sources [10]. The decisions can be constructed using the different fusion schemes and it is necessary to compare the potentially achievable classification accuracies in these schemes.

© Springer Nature Switzerland AG 2019
V. V. Strijov et al. (Eds.): IDP 2016, CCIS 794, pp. 28–43, 2019.
https://doi.org/10.1007/978-3-030-35400-8_3

The classification problem in the ensemble of sources is similar to the source coding problem based on quantization [8]. For blocks of the continuous values, there are known the schemes of scalar and vector quantization. The scalar scheme uses an independent one-dimensional quantization for the individual values while the vector scheme uses a multi-dimensional quantization for blocks of the values. Both quantization schemes transform the blocks of the continuous values into the appropriate blocks of the quantized values that yield the code vectors.

It should be noted that the vector quantization is constructed with covering the multi-dimensional space of the sources values by general spheres [9]. In scalar quantization, the same multi-dimensional space is covered by cubes, whose edge size yields a quantization step in any dimension. The centers of the above spheres or cubes give the code vectors for the submitted blocks of the continuous values. Since, for a fixed volume, the compactness of the spheres relative to cubes is higher, the vector quantization provides a smaller error with respect to the scalar quantization. Thus, for the fixed number of the covering spheres or cubes, the fusion of the continuous values into blocks before vector quantization provides the more efficient code vectors as against the code vectors to be yielded by the fusion of the scalar quantized values.

Also, for classification in a given ensemble of the sources, a smaller error probability is waited in a scheme of joint classifying each composite object as compared to an error probability in the scheme of combining the decisions on the objects of the individual sources. The proposed paper is focused on both developing a theoretical validity of this idea and supporting the main result by a computing experiment.

We investigate two fusion schemes that use the different data compositions for making the decisions on the composite objects from the ensemble of the sources. They are WMV scheme by weighted majority voting the decisions on the objects of the individual sources [12] and GDM scheme, which combines the sources with a general dissimilarity measure between any pare of the composite objects [15]. Notice that WMV scheme is based on a composition of the soft decisions on the objects of individual sources while GDM scheme uses a composition of metrics in datasets of the sources. Thus, ideologically, WMV and GDM fusion schemes are similar to the above scalar and vector quantization.

Some limits on the majority vote accuracy have been obtained in [11]. Also, it is clear that the multiclass classification error probability depends on the average mutual information [6] between a set of the objects and a set of the classes. Moreover, the larger average mutual information should provide the smaller error probability. So, our goal is to introduce the mutual information-based characteristics for WMV and GDM fusion schemes and to show an advantage of GDM scheme relative to WMV scheme in the error probability.

2   Formalization of the Problem

2.1   Basic Definitions and Classification Schemes

Let $\Omega = \{\omega_1, \ldots, \omega_c\}$, $c \geq 2$, be a set of classes of the prior probabilities $P(\omega_i) > 0 : \sum_{i=1}^{c} P(\omega_i) = 1$, and $\mathbf{X}^M = \mathbf{X}_1, \ldots, \mathbf{X}_M$ be an ensemble of sources, where the set $\mathbf{X}_m = \{\mathbf{x}_m = (x_{m1}, \ldots, x_{mN_m})\}$, $m = 1, \ldots, M$, of N_m-dimensional vectors gives m-th source objects. In the ensemble $\mathbf{X}^M$, the components of any vector $\mathbf{x}_m \in \mathbf{X}_m$ take real values in $(-\infty, \infty)$, and any composite object $\mathbf{x}^M = (\mathbf{x}_1, \ldots, \mathbf{x}_M) \in \mathbf{X}^M$ is produced by a collection of the vectors by one per source belonging to the same class in Ω.

In each set $\mathbf{X}_m$, $m = 1, \ldots, M$ a dissimilarity measure between any pair of the objects $\mathbf{x}_m \in \mathbf{X}_m$, $\hat{\mathbf{x}}_m \in \mathbf{X}_m$ is defined as

$$d(\mathbf{x}_m, \hat{\mathbf{x}}_m) = \sum_{n=1}^{N_m} \frac{(x_{mn} - \hat{x}_{mn})^2}{\sigma_{mn}^2} \tag{1}$$

where $0 < \sigma_{mn}^2 < \infty$, $n = 1, \ldots, N_m$, are unknown parameters. Also, for any pair of composite objects $\mathbf{x}^M \in \mathbf{X}^M$ and $\hat{\mathbf{x}}^M \in \mathbf{X}^M$, we define a general dissimilarity measure as a weighted composition of the metrics of the form (1) taken with the weights $W = \{w_m > 0, \ m = 1, \ldots, M\}$ as follows

$$D(\mathbf{x}^M, \hat{\mathbf{x}}^M) = \sum_{m=1}^{M} w_m d(\mathbf{x}_m, \hat{\mathbf{x}}_m). \tag{2}$$

Let

$$\{\mathbf{x}_{im}, \ i = 1, \ldots, c\} \subset \mathbf{X}_m; \ m = 1, \ldots, M \tag{3}$$

be the subsets of the source template objects that represent the classes with one object from $\mathbf{X}_m$ per each class. The subsets (3) produce the subset of the template composite objects

$$\{\mathbf{x}_i^M = (\mathbf{x}_{i1}, \ldots, \mathbf{x}_{iM}), \ i = 1, \ldots, c\} \subset \mathbf{X}^M \tag{4}$$

Using the dissimilarity measure (1) and assuming a compactness of the objects in $\mathbf{X}_m$, $m = 1, \ldots, M$ relative to the corresponding template objects in (3), we define class-conditional densities

$$p(\mathbf{x}_m | \omega_i) = \frac{e^{-d(\mathbf{x}_m, \mathbf{x}_{im})}}{\int_{\mathbf{X}_m} e^{-d(\mathbf{x}_m, \mathbf{x}_{im})} d\mathbf{x}_m}, \ i = 1, \ldots, c. \tag{5}$$

Similarly, assuming a compactness of the composite objects in $\mathbf{X}^M$ relative to the corresponding templates in (4) and using the general dissimilarity measure (2), we define class-conditional densities

$$p_W(\mathbf{x}^M | \omega_i) = \frac{e^{-D(\mathbf{x}^M, \mathbf{x}_i^M)}}{\int_{\mathbf{X}^M} e^{-D(\mathbf{x}^M, \mathbf{x}_i^M)} d\mathbf{x}^M} = \prod_{m=1}^{M} \frac{e^{-w_m d(\mathbf{x}_m, \mathbf{x}_{im})}}{\int_{\mathbf{X}_m} e^{-w_m d(\mathbf{x}_m, \mathbf{x}_{im})} d\mathbf{x}_m}, i = 1, \ldots, c. \tag{6}$$

Under the product in (6), the weighted class-conditional densities

$$p_{w_m}(\mathbf{x}_m|\omega_i) = \frac{e^{-w_m d(\mathbf{x}_m,\mathbf{x}_{im})}}{\int_{\mathbf{X}_m} e^{-w_m d(\mathbf{x}_m,\mathbf{x}_{im})}d\mathbf{x}_m}, \; i = 1,\ldots,c \qquad (7)$$

yield the class-conditional densities of the form (5) when $w_m = 1$. In terms of information theory, the densities (7) define the m-th source observation channel between input set Ω and the output set $\mathbf{X}_m$ as well as the densities (6) yield the observation multi-channel between Ω and $\mathbf{X}^M$. In WMV scheme, the source observation channels are taken with the unit weights $w_m = 1$, $m = 1,\ldots,M$ whereas GDM scheme utilizes the different positive weights $w_m > 0$, $m = 1,\ldots,M$ in the corresponding source observation channels.

Next, we introduce the discriminant functions that are needed for making WMV-based and GDM-based decisions on the submitted composite objects. Let

$$g_i^d(\mathbf{x}_m), \; i = 1,\ldots,c$$

be the discriminant functions that are defined in the sets $\mathbf{X}_m$, $m = 1,\ldots,M$ using the dissimilarity measure (1). Calculating these functions for the source components of a submitted composite object $\mathbf{x}^M \in \mathbf{X}^M$, the WMV-based class label decision is defined by

$$j^{\text{WMV}}(\mathbf{x}^M) = \arg\max_{i=1}^c \sum_{m=1}^M w_m g_i^d(\mathbf{x}_m). \qquad (8)$$

The sum in (8) is taken with the positive source weights and the discriminant functions are independent on the weights. Similarly, using in the ensemble $\mathbf{X}^M$ the discriminant functions

$$g_i^D(\mathbf{x}^M), \; i = 1,\ldots,c$$

that are dependent on the weights W, we obtain the GDM-based class label decision on the same $\mathbf{x}^M \in \mathbf{X}^M$ as follows

$$j^{\text{GDM}}(\mathbf{x}^M) = \arg\max_{i=1}^c g_i^D(\mathbf{x}^M). \qquad (9)$$

The schemes of the classifiers by the decision rules (8) and (9) are shown in Fig. 1. Here, $\hat{\Omega} = \Omega$ provided that the decisions in $\hat{\Omega}$ can differ from the real classes in Ω. The prior distribution of the classes and the observation channels for all sources allow us to compare WMV and GDM fusion schemes in terms of an information criterion based on the values of the average mutual information between $\mathbf{X}_m$, $m = 1,\ldots,M$ and Ω.

2.2 Information Criterion of Efficiency for WMV and GDM Fusion Schemes

Given the prior distribution $\{P(\omega_i)\}$ and the weighted class-conditional densities $\{p_{w_m}(\mathbf{x}_m|\omega_i)\}$ of the form (7), $i = 1,\ldots,c$, the average mutual information

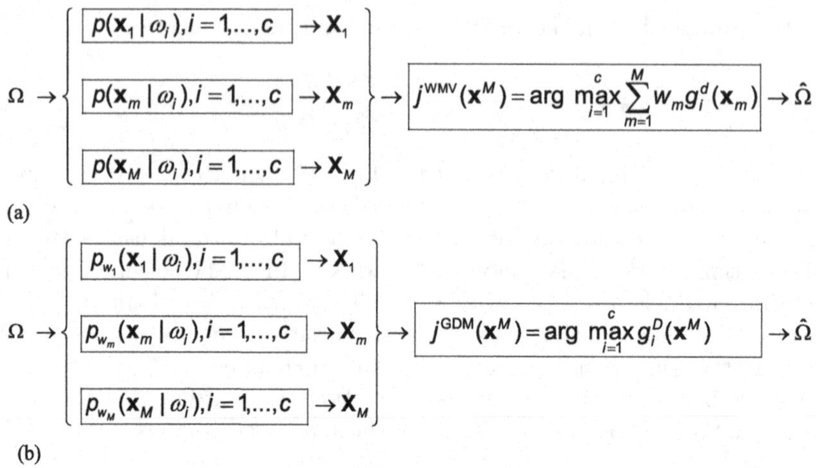

Fig. 1. Schemes of WMV-based (a) and GDM-based (b) classifiers

between $\mathbf{X}_m$ and Ω is defined according to [6] by

$$I_{w_m}(\mathbf{X}_m;\Omega) = H_{w_m}(\mathbf{X}_m) - H_{w_m}(\mathbf{X}_m|\Omega). \tag{10}$$

Here,

$$H_{w_m}(\mathbf{X}_m) = -\int_{\mathbf{X}_m} p_{w_m}(\mathbf{x}_m)\ln p_{w_m}(\mathbf{x}_m)\mathrm{d}\mathbf{x}_m$$

and

$$H_{w_m}(\mathbf{X}_m|\Omega) = -\sum_{i=1}^{c} P(\omega_i)\int_{\mathbf{X}_m} p_{w_m}(\mathbf{x}_m|\omega_i)\ln p_{w_m}(\mathbf{x}_m|\omega_i)\mathrm{d}\mathbf{x}_m$$

are the differential entropies, and $p_{w_m}(\mathbf{x}_m) = \sum_{i=1}^{c} P(\omega_i)p_{w_m}(\mathbf{x}_m|\omega_i)$ is the marginal density in $\mathbf{X}_m$, $m = 1,\ldots,M$. Notice that, for any $w_m > 0$, the average mutual information in (10) is nonnegative and it does not exceed the entropy

$$H(\Omega) = -\sum_{i=1}^{c} P(\omega_i)\ln P(\omega_i).$$

At the point $w_m = 1$, $p_{w_m}(\mathbf{x}_m|\omega_i) = p(\mathbf{x}_m|\omega_i)$ and $I_{w_m}(\mathbf{X}_m;\Omega) = I(\mathbf{X}_m;\Omega)$.

Taking the means of the values $I(\mathbf{X}_m;\Omega)$ and $I_{w_m}(\mathbf{X}_m;\Omega)$ over all of the sources in the ensemble, we obtain the efficiency characteristics for the classifiers by the decisions (8) and (9), respectively. These characteristics are defined as follows

$$I_{W\_\mathrm{mean}}^{\mathrm{WMV}}(\mathbf{X}^M;\Omega) = \sum_{m=1}^{M} I(\mathbf{X}_m;\Omega)\frac{w_m}{\sum_{m=1}^{M} w_m}; \tag{11}$$

$$I_{W\_\mathrm{mean}}^{\mathrm{GDM}}(\mathbf{X}^M;\Omega) = \frac{1}{M}\sum_{m=1}^{M} I_{w_m}(\mathbf{X}_m;\Omega). \tag{12}$$

Our goal is to prove the inequality

$$\max_W I_{W\_\text{mean}}^{\text{WMV}}(\mathbf{X}^M;\Omega) \leq I_{W^*\_\text{mean}}^{\text{GDM}}(\mathbf{X}^M;\Omega) \tag{13}$$

where W^* is the collection of the weights giving the maximum.

2.3 Average Mutual Information and Classification Error Probability

The criterion (13) is based on a dependence of the average mutual information $I(\mathbf{X}^M;\hat{\Omega})$ between the ensemble $\mathbf{X}^M$ and the set of the decisions $\hat{\Omega}$ on a potentially attainable error probability ε in the classification schemes shown in Fig. 1. Given observation multi-channel, such function has been defined in [14] as a generalization of the rate-distortion function for the source coding model with a noisy observation channel [3]. According to [14], the above function is lower bounded by

$$R_L(\varepsilon) = I(\mathbf{X}^M;\Omega) - h(\varepsilon - \varepsilon_{\min}) - (\varepsilon - \varepsilon_{\min})\ln(c-1), \varepsilon_{\min} \leq \varepsilon \leq \varepsilon_{\max}. \tag{14}$$

Here,

$$h(z) = -z\ln z - (1-z)\ln(1-z); \quad R_L(\varepsilon_{\min}) = I(\mathbf{X}^M;\Omega); \quad R_L(\varepsilon_{\max}) = 0$$

and $I(\mathbf{X}^M;\Omega) = H(\mathbf{X}^M) - H(\mathbf{X}^M|\Omega)$ is the average mutual information between the input and the output of the observation multi-channel shown in Fig. 1. The function (14) decreases when ε increases and takes the largest value $I(\mathbf{X}^M;\Omega)$ at the point $\varepsilon_{\min}$. It is not difficult to show that the minimal error rate $\varepsilon_{\min}$ is lower estimated by the conditional entropy $H(\Omega|\mathbf{X}^M)$ and $\varepsilon_{\min}$ tends to the zero when $H(\Omega|\mathbf{X}^M)$ decreases. Moreover, an increase of the entropy $H(\Omega|\mathbf{X}^M)$ can be performed by increasing the size M of the ensemble. Taking into account the symmetry of the average mutual information $I(\mathbf{X}^M;\Omega) = H(\mathbf{X}^M) - H(\mathbf{X}^M|\Omega) = H(\Omega) - H(\Omega|\mathbf{X}^M)$, the bound (14) in the limit yields the Shannon bound of the form $H(\Omega) - h(\varepsilon) - \varepsilon\ln(c-1)$ [6].

In (14), the average mutual information $I(\mathbf{X}^M;\Omega)$ is defined in the product $\Omega * \mathbf{X}^M$ using the prior probabilities of the classes and the class-conditional densities of the form (6). According to Fig. 1, the class-conditional densities in GDM scheme depend on the source weights and therefore $I(\mathbf{X}^M;\Omega) = I_W^{\text{GDM}}(\mathbf{X}^M;\Omega)$ is a function of W. In WMV scheme, the corresponding average mutual information $I(\mathbf{X}^M;\Omega) = I_W^{\text{WMV}}(\mathbf{X}^M;\Omega)$ is equal to $I_W^{\text{GDM}}(\mathbf{X}^M;\Omega)$ taken with the weights $w_m = 1, m = 1,\ldots,M$. The values $I^{\text{WMV}}(\mathbf{X}^M;\Omega)$ and $I_W^{\text{GDM}}(\mathbf{X}^M;\Omega)$ correspond to the minimal error probabilities $\varepsilon_{\min}^{\text{WMV}}$ and $\varepsilon_{\min}^{\text{GDM}}$ in (14) that are defined by the conditional entropy $H(\Omega|\mathbf{X}^M)$ in WMV and GDM fusion schemes, respectively. In both fusion schemes, the minimal error probabilities can be achieved by the Bayes decisions of the form (9) with the discriminant functions given by the posterior probabilities [4].

In general, the source datasets $\mathbf{X}_1,\ldots,\mathbf{X}_M$ are statistically dependent and there are valid the relations

$$I_{W\_\text{mean}}^{\text{WMV}}(\mathbf{X}^M;\Omega) < I^{\text{WMV}}(\mathbf{X}^M;\Omega);$$

$$I_{W\_\text{mean}}^{\text{GDM}}(\mathbf{X}^M;\Omega) < I_W^{\text{GDM}}(\mathbf{X}^M;\Omega).$$

For the weights W^* giving the maximum in (13), the means $I_{W^*\_\text{mean}}^{\text{WMV}}(\mathbf{X}^M;\Omega)$ and $I_{W^*\_\text{mean}}^{\text{GDM}}(\mathbf{X}^M;\Omega)$ yield the error probabilities $\varepsilon^{\text{WMV}} > \varepsilon_{\min}^{\text{WMV}}$ and $\varepsilon^{\text{GDM}} > \varepsilon_{\min}^{\text{GDM}}$ that belong to the corresponding bounds of the form (14). Also, taking into account that $I^{\text{WMV}}(\mathbf{X}^M;\Omega) \leq I_{W^*}^{\text{GDM}}(\mathbf{X}^M;\Omega)$, the inequality (13) provides the relation between the potentially achievable error probabilities as follows $\varepsilon^{\text{WMV}} \geq \varepsilon^{\text{GDM}}$. Figure 2 illustrates this fact. Notice that these error probabilities are the lower estimated values to be independent on the constructive decision algorithms.

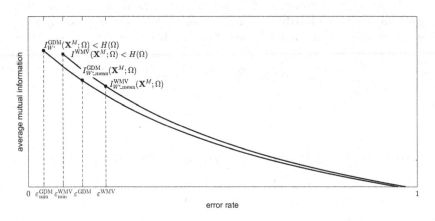

Fig. 2. Sketches of the lower bounds to the average mutual information as the function of the error probability in WMV and GDM fusion schemes

3   Calculation of the Average Mutual Information

In this section, an upper estimate for the mutual information $I_{w_m}(\mathbf{X}_m;\Omega)$ in (10) is obtained as a function of the variable $w_m^{1/2}$. At the point $w_m^{1/2} = 1$, this function yields the upper estimate for $I(\mathbf{X}_m;\Omega)$. Using the marginal density $p_{w_m}(\mathbf{x}_m)$ and taking into account that $-\ln z$ is the convex downwards function of z, it is valid the following Jensen inequality [2]

$$-\ln p_{w_m}(\mathbf{x}_m) = -\ln \sum_{i=1}^c P(\omega_i)p_{w_m}(\mathbf{x}_m|\omega_i) \leq -\sum_{i=1}^c P(\omega_i)\ln p_{w_m}(\mathbf{x}_m|\omega_i).$$

Applying this inequality for $H_{w_m}(\mathbf{X}_m)$ in (10), we obtain the upper estimated differential entropy as follows

$$H_{w_m}(\mathbf{X}_m) \leq -\sum_{i=1}^c P(\omega_i)\sum_{j=1}^c P(\omega_j)\int_{\mathbf{X}_m} p_{w_m}(\mathbf{x}_m|\omega_i)\ln p_{w_m}(\mathbf{x}_m|\omega_j)\mathrm{d}\mathbf{x}_m. \quad (15)$$

Given the dissimilarity measures (1) and (2), the conditional density $p_{w_m}(\mathbf{x}_m|\omega_i)$ in (7) is the Gaussian density for N_m independent variables with the means x_{imn} and the variances $\sigma^2_{imn}/(2w_m)$, $n = 1,\ldots,N_m$ subject to $w_m > 0$. It amounts the integral in (15) over the area $(-\infty, +\infty)$ to the Euler integrals [7]. The calculations yield the upper estimated differential entropy

$$
\begin{aligned}
H_{w_m}(\mathbf{X}_m) \leq \tfrac{1}{2}\ln\tfrac{\pi}{w_m} + \tfrac{1}{2}\sum_{j=1}^{c} P(\omega_j)\sum_{n=1}^{N_m}\ln\sigma^2_{jmn} \\
+ w_m\sum_{i=1}^{c} P(\omega_i)\sum_{j=1}^{c} P(\omega_j)\sum_{n=1}^{N_m}\frac{(x_{imn}-x_{jmn})^2}{\sigma^2_{jmn}} \\
+ 2\frac{w_m^{1/2}}{\sqrt{\pi}}\sum_{i=1}^{c} P(\omega_i)\sum_{j=1}^{c} P(\omega_j)\sum_{n=1}^{N_m}\frac{|x_{imn}-x_{jmn}|\sigma_{imn}}{\sigma^2_{jmn}} \\
+ \tfrac{1}{2}\sum_{i=1}^{c} P(\omega_i)\sum_{j=1}^{c} P(\omega_j)\sum_{n=1}^{N_m}\frac{\sigma^2_{imn}}{\sigma^2_{jmn}}.
\end{aligned}
\tag{16}
$$

and the conditional differential entropy

$$
H_{w_m}(\mathbf{X}_m|\Omega) = \tfrac{1}{2}\ln\tfrac{\pi e}{w_m} + \tfrac{1}{2}\sum_{i=1}^{c} P(\omega_i)\sum_{n=1}^{N_m}\ln\sigma^2_{imn}.
\tag{17}
$$

The substitutions in (10) by the corresponding parts in (16) and (17) give the upper estimated average mutual information as follows

$$
\begin{aligned}
I_{w_m}(\mathbf{X}_m;\Omega) \leq w_m\sum_{i=1}^{c} P(\omega_i)\sum_{j=1}^{c} P(\omega_j)\sum_{n=1}^{N_m}\frac{(x_{imn}-x_{jmn})^2}{\sigma^2_{jmn}} \\
+ 2\frac{w_m^{1/2}}{\sqrt{\pi}}\sum_{i=1}^{c} P(\omega_i)\sum_{j=1}^{c} P(\omega_j)\sum_{n=1}^{N_m}\frac{|x_{imn}-x_{jmn}|\sigma_{imn}}{\sigma^2_{jmn}} \\
+ \tfrac{1}{2}\sum_{i=1}^{c} P(\omega_i)\sum_{j=1}^{c} P(\omega_j)\sum_{n=1}^{N_m}\left(\frac{\sigma^2_{imn}}{\sigma^2_{jmn}} - 1\right).
\end{aligned}
\tag{18}
$$

The right part in (18) is a parabolic function $a_m w_m + b_m w_m^{1/2} + c_m$ of the variable $w_m^{1/2} > 0$, $m = 1,\ldots,M$. Here, the coefficients $a_m > 0$, $b_m > 0$ and $c_m \geq 0$ depend on the template objects in classes as well as on parameters of the dissimilarity measure (1) and the prior probabilities of the classes. Notice that a positivity of a_m and b_m is caused by the distinction of the templates in different classes. A non-negativity of c_m follows from the Jensen inequality

$$
\sum_{i=1}^{c} P(\omega_i)\sum_{j=1}^{c} P(\omega_j)\frac{\sigma^2_{imn}}{\sigma^2_{jmn}} \geq \frac{\sum_{i=1}^{c} P(\omega_i)\sigma^2_{imn}}{\sum_{j=1}^{c} P(\omega_j)\sigma^2_{jmn}} = 1.
$$

At the point $w_m^{1/2} = 1$, the right part in (18) yields the upper estimate $a_m + b_m + c_m$ for $I(\mathbf{X}_m;\Omega)$. The sketch of the parabola $a_m w_m + b_m w_m^{1/2} + c_m$ is shown in Fig. 3 by the solid branch, which increases and exceeds $c_m \geq 0$ for $w_m^{1/2} > 0$. Also, the weights of interest are defined by the values $w_m^{1/2} > 1$ that satisfy the condition $a_m w_m + b_m w_m^{1/2} + c_m \leq H(\Omega)$, $m = 1, \ldots, M$.

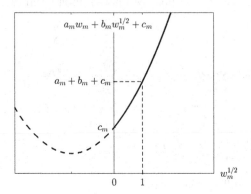

Fig. 3. The sketch of $a_m w_m + b_m w_m^{1/2} + c_m$ as a function of the variable $w_m^{1/2} > 0$

Setting $\delta_m = (a_m + b_m + c_m)/H(\Omega) \leq 1$, we assign the source weights

$$w_m(s) = e^{s\delta_m}, \ m = 1, \ldots, M \tag{19}$$

where $s \geq 0$ is a free parameter that provides $w_m(s) \geq 1$. In what follows, the upper estimates (18) with the weights (19) are denoted by $I_s(\mathbf{X}_m; \Omega)$, $m = 1, \ldots, M$.

4   Main Results

Using in (11) the estimates $I(\mathbf{X}_m; \Omega) \leq a_m + b_m + c_m$, $m = 1, \ldots, M$ and the source weights (19), we obtain the upper estimated mean value $I_{s\_mean}^{WMV}(\mathbf{X}^M; \Omega)$. The estimates $I_{w_m}(\mathbf{X}_m; \Omega) \leq a_m w_m + b_m w_m^{1/2} + c_m$, $m = 1, \ldots, M$ taken with the similar weights in (12) yield the upper estimated mean value $I_{s\_mean}^{GDM}(\mathbf{X}^M; \Omega)$. Then for $s \to 0$, we calculate the asymptotic maximum $I_{s^*\_mean}^{WMV}(\mathbf{X}^M; \Omega)$ at the point s^* that gives the inequality $I_{s^*\_mean}^{WMV}(\mathbf{X}^M; \Omega) \leq I_{s^*\_mean}^{GDM}(\mathbf{X}^M; \Omega)$.

For a compactness of the subsequent formulas, we use the following notations

$$\mu = \frac{1}{M} \sum_{m=1}^{M} \delta_m;$$

$$\Delta_1 = \frac{1}{M} \sum_{m=1}^{M} \delta_m^2 - \mu^2;$$

$$\Delta_2 = \frac{1}{M} \sum_{m=1}^{M} \delta_m^3 - \mu \frac{1}{M} \sum_{m=1}^{M} \delta_m^2.$$

Theorem 1. *For $2\mu\Delta_1 - \Delta_2 > \Delta_1 > 0$ and $s \to 0$, the value $s^* = \Delta_1/(2\mu\Delta_1 - \Delta_2)$ yields*

$$\max_s I_{s\_mean}^{WMV}(\mathbf{X}^M; \Omega) = (\mu + \frac{1}{2}\Delta_1 s^*)H(\Omega).$$

For $\Delta_1 = 0$ and $s \geq 0$, there is valid the equality $I_{s\_mean}^{WMV}(\mathbf{X}^M; \Omega) = \mu H(\Omega)$.

Proof. Using $q_s(\delta_m) = e^{s\delta_m}/\sum_{m=1}^M e^{s\delta_m}$, the upper estimated mean value (11) takes the form

$$I_{s\_mean}^{WMV}(\mathbf{X}^M; \Omega) = H(\Omega) \sum_{m=1}^M \delta_m q_s(\delta_m). \tag{20}$$

For $s \to 0$, there is valid the asymptotic equation

$$\sum_{m=1}^M \delta_m q_s(\delta_m) \approx \mu + \Delta_1 s - \frac{1}{2}(2\mu\Delta_1 - \Delta_2)s^2. \tag{21}$$

In case of $\Delta_1 > 0$, the parabola in the right part of (21) takes the maximal value $\mu + \Delta_1 s^*/2$ at the point $s^* = \Delta_1/(2\mu\Delta_1 - \Delta_2)$. Notice that the same values $\delta_m = \delta$, $m = 1,\ldots,M$ provide $\Delta_1 = 0$ and $\Delta_2 = 0$. In this case $q_s(\delta_m) = 1/M$, and the sum in (21) is equal to $\mu = \delta$ for all $s \geq 0$. The substitution of the sum in (20) by $\mu + \Delta_1 s^*/2$ in case of $\Delta_1 > 0$ or by μ in case of $\Delta_1 = 0$ completes the proof. $\square$

Theorem 2. *For $\Delta_1 > 0$ and on condition that $a_m \geq c_m$, $m = 1,\ldots,M$, there is valid the inequality $I_{s^*\_mean}^{WMV}(\mathbf{X}^M; \Omega) < I_{s^*\_mean}^{GDM}(\mathbf{X}^M; \Omega)$ at the point $s^* > 0$. For $\Delta_1 = 0$ and $s \geq 0$, there is valid the inequality $I_{s\_mean}^{WMV}(\mathbf{X}^M; \Omega) < I_{s\_mean}^{GDM}(\mathbf{X}^M; \Omega)$, which passes into the equality at the point $s = 0$.*

Proof. The estimates (18) taken with the weights (19) give the upper estimated mean value (12) as follows

$$I_{s\_mean}^{GDM}(\mathbf{X}^M; \Omega) = \frac{1}{M} \sum_{m=1}^M \left(a_m e^{s\delta_m} + b_m e^{s\delta_m/2} + c_m \right). \tag{22}$$

Taking the square approximations of the exponential terms in (22), we obtain the following inequality

$$I_{s\_mean}^{GDM}(\mathbf{X}^M; \Omega) \geq \mu H(\Omega) + \left(\frac{1}{M}\sum_{m=1}^M a_m\delta_m + \frac{1}{2M}\sum_{m=1}^M b_m\delta_m \right) s$$
$$+ \left(\frac{1}{2M}\sum_{m=1}^M a_m\delta_m^2 + \frac{1}{4M}\sum_{m=1}^M b_m\delta_m^2 \right) s^2. \tag{23}$$

In case of $\Delta_1 > 0$, the inequality (23) together with the estimates (20) and (21) yield

$$I_{s\_mean}^{GDM}(\mathbf{X}^M; \Omega) - I_{s\_mean}^{WMV}(\mathbf{X}^M; \Omega) \geq (\sum_{m=1}^M \frac{a_m\delta_m}{M} + \sum_{m=1}^M \frac{b_m\delta_m}{2M} - \Delta_1 H(\Omega))s$$
$$+ (\sum_{m=1}^M \frac{a_m\delta_m^2}{2M} + \sum_{m=1}^M \frac{b_m\delta_m^2}{4M} + \frac{1}{2}(2\mu\Delta_1 - \Delta_2)H(\Omega))s^2. \tag{24}$$

Under the assumption

$$\frac{1}{M} \sum_{m=1}^{M} a_m \delta_m + \frac{1}{2M} \sum_{m=1}^{M} b_m \delta_m \geq \frac{1}{2} \Delta_1 H(\Omega) \tag{25}$$

the right part in (24) is lower estimated by the parabola

$$-\frac{1}{2} \Delta_1 H(\Omega) s + \left(\frac{1}{2M} \sum_{m=1}^{M} a_m \delta_m^2 + \frac{1}{4M} \sum_{m=1}^{M} b_m \delta_m^2 + \frac{1}{2} (2\mu\Delta_1 - \Delta_2) H(\Omega) \right) s^2$$

that has a positive root

$$s_0 = \frac{\Delta_1 H(\Omega)}{(2\mu\Delta_1 - \Delta_2) H(\Omega) + \frac{1}{M} \sum_{m=1}^{M} a_m \delta_m^2 + \frac{1}{2M} \sum_{m=1}^{M} b_m \delta_m^2} < \frac{\Delta_1}{2\mu\Delta_1 - \Delta_2} = s^*.$$

Since the parabola is positive for $s > s_0$, the lower estimate for the right part in (24) is positive at the point $s^* > 0$ of the maximal value $I_{s^*\_\text{mean}}^{\text{WMV}}(\mathbf{X}^M; \Omega)$ that provides the inequality $I_{s^*\_\text{mean}}^{\text{GDM}}(\mathbf{X}^M; \Omega) - I_{s^*\_\text{mean}}^{\text{WMV}}(\mathbf{X}^M; \Omega) > 0$.

Notice that the assumption (25) is equivalent to the inequality

$$\frac{1}{M} \sum_{m=1}^{M} (c_m - a_m) \delta_m \leq \mu^2 H(\Omega)$$

which is valid on the theorem condition $a_m \geq c_m$, $m = 1, \ldots, M$. This condition is held if the templates in different classes are sufficiently distinct from each other. Formally, the parameters in (18) should satisfy the following inequalities

$$(x_{imn} - x_{jmn})^2 \geq \frac{1}{2} |\sigma_{imn}^2 - \sigma_{jmn}^2|, \ m = 1, \ldots, M, \ n = 1, \ldots, N_m.$$

In case of $\Delta_1 = 0$, we have

$$I_{s\_\text{mean}}^{\text{WMV}}(\mathbf{X}^M; \Omega) = \mu H(\Omega);$$

$$I_{s\_\text{mean}}^{\text{GDM}}(\mathbf{X}^M; \Omega) \geq \mu H(\Omega)$$

for $s \geq 0$. Thus, there is valid the inequality

$$I_{s\_\text{mean}}^{\text{WMV}}(\mathbf{X}^M; \Omega) \leq I_{s\_\text{mean}}^{\text{GDM}}(\mathbf{X}^M; \Omega)$$

which passes into the equality at the point $s = 0$. The theorem is proved.   □

Sketches of the graphics in Fig. 4 illustrate the Theorems 1 and 2.

Corollary 1. *For WMV and GDM fusion schemes, the parameters s^* in case of $\Delta_1 > 0$ and $s > 0$ in case of $\Delta_1 = 0$ give the means of the average mutual information per one source that provide the lower bounds to the appropriate error probabilities satisfying the inequality $\varepsilon^{\text{WMV}} > \varepsilon^{\text{GDM}}$.*

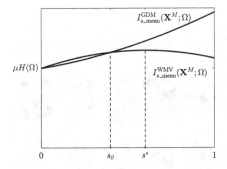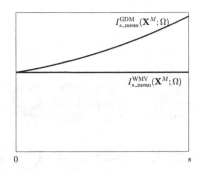

Fig. 4. The graphical illustration of the results in cases of $\Delta_1 > 0$ (left) and $\Delta_1 = 0$ (right)

5 Experimental Results

We evaluate the efficiency of WMV and GDM fusion schemes by their error rates for face recognition of HSI color images. The components H, S, and I give the objects of the individual sources and the ensemble HSI produces the composite objects. The color images are taken from 25 persons (classes) per 40 images in each class [1]. The prior probability distribution of the classes is uniform. Face recognition has been performed in a space of multilevel tree-structured pattern representations with elliptic primitives [15]. The error rates have been obtained for multiclass NN (nearest neighbor) and SVM (support vector machine) classifiers that are the collections of elementary "class-vs-all" classifiers. The experiments have been performed using 100 times, 2 fold cross-validation.

Examples of the tree-structured representations for the face images that are produced by the sources H, S and I are shown in Fig. 5. An object (pattern) $\mathbf{x}$ of any source is represented in a form of a sequence of the representations $\mathbf{x}^L = (x_0, \ldots, x_l, \ldots, x_L)$. Here, $L = 8$ and x_l is a collection of the elliptic primitives belonging to all end nodes in the subtree of depth l including the levels $0, \ldots, l$. Any representation $\mathbf{x}^L$ is constructed by dichotomic partitioning the pattern $\mathbf{x} \in \mathbf{X}$ into the segments and approximating each n-th segment by the elliptic primitive Q_n, which corresponds to the n-th node in the binary tree. Any partitioned n-th segment produces a pair of the new segments of the numbers $2n + 1$ and $2n + 2$. The root primitive Q_0 approximates the source pattern $\mathbf{x}$, which is the zero segment. Each elliptic primitive $Q_n = (\mathbf{r}_n, \mathbf{u}_n, \mathbf{v}_n, z_n)$ of the number n is calculated in the principal axes of the corresponding segment and the primitive is defined by a vector of the segment center $\mathbf{r}_n$, the vectors of semiaxes $\mathbf{u}_n$ and $\mathbf{v}_n$, and a value of the mean darkness z_n over the pixels in the approximated segment. The parameters of each primitive Q_n are normalized by the corresponding parameters of the root primitive Q_0. Thus, the representation $\mathbf{x}^L$ is approximately invariant to rotation, translation, scale, and darkness of the pattern $\mathbf{x}$.

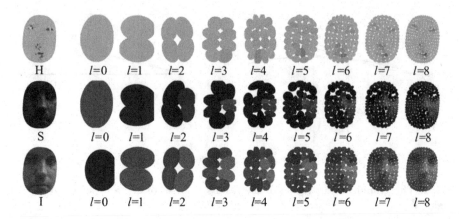

Fig. 5. Examples of the 8th level representations for the face HSI images

For any source H, S, or I, the dissimilarity measure between a pair of the source objects $\mathbf{x}$ and $\hat{\mathbf{x}}$ can be calculated using the dissimilarities of the same number primitives Q_n and $\hat{Q}_n$ that belong to the corresponding representations $\mathbf{x}^L$ and $\hat{\mathbf{x}}^L$. Given ensemble of size $M \geq 2$, for the m-th source, the dissimilarity measure has been defined in [13] by

$$d(\mathbf{x}_m, \hat{\mathbf{x}}_m) = d_r(\mathbf{x}_m, \hat{\mathbf{x}}_m) + d_{uv}(\mathbf{x}_m, \hat{\mathbf{x}}_m) + d_z(\mathbf{x}_m, \hat{\mathbf{x}}_m)$$

where the components give the dissimilarities of the corresponding elliptic primitives in their centers, semiaxes, and darkness values, respectively. The weighted sum of the above measures over M sources yields the general dissimilarity measure $D(\mathbf{x}_m, \hat{\mathbf{x}}_m)$ of the form (2) between the composite objects $\mathbf{x}^M$ and $\hat{\mathbf{x}}^M$.

The discriminant functions of the elementary NN and SVM classifiers are constructed at the learning stage using the source train sets

$$\mathbf{X}_m^{\text{train}} \subset \mathbf{X}_m, \; m = 1, \ldots, M$$

and the ensemble train set

$$\mathbf{X}^{M\text{-train}} = \mathbf{X}_1^{\text{train}}, \ldots, \mathbf{X}_M^{\text{train}}$$

of the same cardinality $K = \|\mathbf{X}^{M\text{-train}}\| = \|\mathbf{X}_m^{\text{train}}\|$, $m = 1, \ldots, M$. The above train sets are divided into the clusters so that

$$\mathbf{X}_m^{\text{train}} = \bigcup_{i=1}^{c} \mathbf{X}_{im}^{\text{train}}, \; m = 1, \ldots, M$$

and

$$\mathbf{X}_i^{M\text{-train}} = \mathbf{X}_{i1}^{\text{train}} \ldots \mathbf{X}_{iM}^{\text{train}}, \; i = 1, \ldots, c.$$

For the i-th elementary NN classifier, the discriminant function $g_i^d(\mathbf{x}_m)$ and $g_i^D(\mathbf{x}^M)$ are calculated using the distances by the corresponding dissimilarity measure between $\mathbf{x}_m \in \mathbf{X}_m$ or $\mathbf{x}^M \in \mathbf{X}^M$ and the nearest neighbor in collection of templates $\hat{\mathbf{X}}_{im} \subset \hat{\mathbf{X}}_{im}^{train}$ or $\hat{\mathbf{X}}_i^M \subset \hat{\mathbf{X}}_i^{M\_train}$, respectively. The discriminant functions, for the i-th SVM classifiers, are calculated in the linear spaces of secondary features [5]. In such space, any source object $\mathbf{x}_m \in \mathbf{X}_m$ is transformed into a vector of the distances between $\mathbf{x}_m$ and all template objects in the collection $\hat{\mathbf{X}}_m = \bigcup\limits_{i=1}^{c} \hat{\mathbf{X}}_{im}$ as follows

$$\mathbf{x}_m \to \mathbf{y}^d(\mathbf{x}_m) = d(\mathbf{x}_m, \hat{\mathbf{x}}_{m1}), \ldots, d(\mathbf{x}_m, \hat{\mathbf{x}}_{mK}),\ \hat{\mathbf{x}}_{mk} \in \hat{\mathbf{X}}_m,\ k = 1, \ldots, K.$$

Similarly, any composite object $\mathbf{x}^M \in \mathbf{X}^M$ is transformed into a vector

$$\mathbf{x}^M \to \mathbf{y}^D(\mathbf{x}^M) = D(\mathbf{x}^M, \hat{\mathbf{x}}_1^M), \ldots, D(\mathbf{x}^M, \hat{\mathbf{x}}_K^M),\ \hat{\mathbf{x}}_k^M \in \hat{\mathbf{X}}^M,\ k = 1, \ldots, K.$$

whose components are the values of the general dissimilarity measure between $\mathbf{x}^M$ and all template composite objects in the set $\hat{\mathbf{X}}^M = \bigcup\limits_{i=1}^{c} \hat{\mathbf{X}}_i^M$ of the same cardinality K. The value of the i-th discriminant function for the vector $\mathbf{y}^d(\mathbf{x}_m)$ or $\mathbf{y}^D(\mathbf{x}^M)$ is calculated using the algebraic distance $\theta_i(\mathbf{y}^d(\mathbf{x}_m))$ or $\theta_i(\mathbf{y}^D(\mathbf{x}^M))$ between the vector and the appropriate i-th separating hyperplane [4]. Thus, for NN and SVM classifiers, the discriminant functions have been defined by the following exponential and sigmoid functions

$$g_i^d(\mathbf{x}_m) = \begin{cases} \exp\{-\min\limits_{\hat{\mathbf{x}}_m \in \mathbf{X}_{im}} d(\mathbf{x}_m, \hat{\mathbf{x}}_m)\}, & \text{for NN} \\ \left(1 + \exp\{-\theta_i(\mathbf{y}^d(\mathbf{x}_m))\}\right)^{-1}, & \text{for SVM} \end{cases}$$

$$g_i^D(\mathbf{x}^M) = \begin{cases} \exp\{-\min\limits_{\hat{\mathbf{x}}^M \in \hat{\mathbf{X}}_i^M} D(\mathbf{x}^M, \hat{\mathbf{x}}^M)\}, & \text{for NN} \\ \left(1 + \exp\{-\theta_i(\mathbf{y}^D(\mathbf{x}^M))\}\right)^{-1}, & \text{for SVM} \end{cases}$$

Notice that the source numbers $m = 1, 2, 3$ correspond to the image components H, S and I.

The Table 1 summarizes the cross-validation error rates for both the individual sources and their ensemble using GDM and WMV fusion schemes. Both fusion schemes provide the smaller error rates in the ensemble HSI relative to the error rates for the sources H, S, and l. Also, GDM scheme shows some advantage as compared with WMV scheme.

Table 1. Error rates for HSI face recognition by NN and SVM classifiers

Classifiers	Sources			Fusion schemes	
	H	**S**	**I**	WMV	GDM
NN	0.022	0.017	0.015	0.009	0.006
SVM	0.019	0.012	0.011	0.007	0.003

Notice that a normalization of the above discriminant functions yields the conditional probabilities for the decisions on a given object. If the discriminant functions to be added by a free parameter, it will provide a possibility to obtain the experimental dependencies of the average mutual information on the error probability for the constructive decision algorithms. These experimental functions can be evaluated for NN and SVM classifiers using the sources H, S, l and the ensemble HSI with WMV and GDM fusion schemes. It will allow us to evaluate a redundancy of the error probability for the constructive classifiers relative to the appropriate lower bounded values given by the function (14).

6   Conclusion

For two fusion schemes in the ensemble of data sources, we have proposed the information characteristics that allow us to compare the potentially achievable classification error probabilities in the investigated schemes. These characteristics are based on the mean values of the average mutual information between the set of the classes and the datasets of the sources and they are independent on the constructive decision algorithm. The mean values of the average mutual information have been defined in WMV fusion scheme that uses a composition of the decisions on the source objects as well as in GDM fusion scheme, which is based on a composition of the metrics in datasets of the sources. Within a given multiclass classification model, we have proved that the information mean value in WMV scheme is smaller to the similar characteristic in GDM scheme. To compare the error probabilities that correspond with the above information characteristics of WMV and GDM schemes, we have used the lower bounds to the rate distortion functions for these schemes. Since the proposed mean values are the points of the appropriate rate distortion functions, we have showed that the lower bound to GDM-based error probability is smaller as against the similar WMV-based error probability. Also, we have performed a computing experiment on face recognition in the ensemble of HSI color images given by the sources H, S, and I. The comparative error rates for NN and SVM classifiers confirm an advantage in accuracy of GDM scheme relative to WMV scheme. In further investigation, we plan to extend the ensemble of biometric sources and the set of the decision algorithms. It is supposed to obtain the experimental dependencies of the average mutual information on the error probability for the sources of faces and signatures as well as for their ensemble using the above fusion schemes. Also, we plan to compare the experimental dependencies with the appropriate lower bounds.

Acknowledgments. The research was supported by the Russian Foundation for Basic Research, the projects 18-07-01231 and 18-07-01385.

References

1. Database of face images. http://sourceforge.net/projects/colorfaces

2. Beckenbach, E., Bellman, R.: Inequalities. Springer, Heidelberg (1961). https://doi.org/10.1007/978-3-642-64971-4

3. Dobrushin, R., Tsybakov, B.: Information transmission with additional noise. IRE Trans. Inf. Theory **8**(5), 293–304 (1962). https://doi.org/10.1109/TIT.1962.1057738

4. Duda, R., Hart, P., Stork, D.: Pattern Classification, 2nd edn. Wiley, Hoboken (2000)

5. Duin, R., de Ridder, D., Tax, D.M.: Experiments with a featureless approach to pattern recognition. Pattern Recogn. Lett. **18**(11–13), 1159–1166 (1997)

6. Gallager, R.: Information Theory and Reliable Communication. Wiley, Hoboken (1968)

7. Gradshteyn, I., Ryzhik, I.: Table of Integrals, Series, and Products, 7th edn. Elsevier, Amsterdam (2007)

8. Gray, R., Neuhoff, D.: Quantization. IEEE Trans. Inf. Theor. **44**(6), 2325–2383 (1998). https://doi.org/10.1109/18.720541

9. Kolmogorov, A., Tikhomirov, V.: ϵ-entropy and ϵ-capacity of sets in functional spaces. In: Shiryayev, A.N. (ed.) Selected Works of A.N. Kolmogorov: Volume III: Information Theory and the Theory of Algorithms, vol. 27, pp. 86–170. Springer, Heidelberg (1993). https://doi.org/10.1007/978-94-017-2973-4_7

10. Kuncheva, L.: Combining Pattern Classifiers: Methods and Algorithms, 2nd edn. Wiley, Hoboken (2014). https://doi.org/10.1002/9781118914564

11. Kuncheva, L., Whitaker, C., Shipp, C., Duin, R.: Limits on the majority vote accuracy in classifier fusion. Pattern Anal. Appl. **6**(1), 22–31 (2003). https://doi.org/10.1007/s10044-002-0173-7

12. Lam, L., Suen, S.Y.: Application of majority voting to pattern recognition: an analysis of its behavior and performance. Trans. Syst. Man Cyber. Part A **27**(5), 553–568 (1997). https://doi.org/10.1109/3468.618255

13. Lange, M., Ganebnykh, S.: On fusion schemes for multiclass object classification with reject in a given ensemble of sources. JPCS **1096**(012048) (2018). https://doi.org/10.1088/1742-6596/1096/1/012048

14. Lange, M., Lange, A.: On information theoretical model for data classification. Mach. Learn. Data Anal. **4**(3), 165–179 (2018). https://doi.org/10.21469/22233792.4.3.03

15. Lange, M., Stepanov, D.: Recognition of objects given by collections of multichannel images. Pattern Recogn. Image Anal. **24**(3), 431–442 (2014). https://doi.org/10.1134/S1054661814030122

On a Metric Kemeny's Median

Sergey Dvoenko[(✉)] and Denis Pshenichny

Tula State University, Tula, Russia
`sergedv@yandex.ru, denispshnechny@yandex.ru`

Abstract. The Kemeny's median represents the coordinated ranking as the opinion of an expert group. Such an opinion is the least different from others in the group and is free of some contradictions (Arrow's paradox) in the well-known majority rule problem. The new problem of building the Kemeny's median with metric characteristics is being developed in this paper. It is assumed, rankings represented by pairwise distances between them are immersed as a set in some Euclidean space. In this case, we can define the mean element as the center of this set. Such central element is a ranking as well and must be similar to the Kemeny's median. The mathematically correct Kemeny's median needs to be seen as the center in its distances to other elements. A new procedure is developed to build the modified loss matrix and find the metric Kemeny's median.

Keywords: Kemeny's median · Rank aggregation · Majority rule · Arrow's paradox · Pairwise comparison · Metrics

1 Introduction

1.1 The Rank Aggregation Problem

In general, ranks aggregation (the collective preference) is the universal problem of discrete optimization. It is known, the aggregation of a linear orders set is NP-hard problem.

Many actual efforts are inspired, for example, by web data retrieval or spam reduction, etc., and focused on developing of special metric binary relations [1], optimized computations based on some additional conditions [2], reformulating to other permutation or some special graph problems [3,4], developing new axiomatics [5,6], geometric analysis [7], etc.

In addition to the known approaches, we suppose that rankings are represented as a set of points in some metric (specifically Euclidean) space. In this case, from one side, we can avoid problems of discrete optimization and we don't need to develop special metric binary relations. From the other side, in the framework of metric approach, we prefer to distinguish between proper metric problems to immerse a set of real scale measurements and a problem to immerse in a metric space just permutations like measurements in (quasi)ordering scales.

The first problem was investigated in [8–10]. Here, we concentrate our attention on the second one.

© Springer Nature Switzerland AG 2019
V. V. Strijov et al. (Eds.): IDP 2016, CCIS 794, pp. 44–57, 2019.
https://doi.org/10.1007/978-3-030-35400-8_4

The well-known choice problem usually consists in selecting from a set of alternatives some element, which supplies some optimal characteristics.

It is recalled, usually it is difficult to prove efficiency of the best choice. The choice, in practice, is often based on expert's intuition and experience. Therefore, we talk about expert's individual choice and their individual preferences. If an expert has successfully selected the best alternative, then, as a rule, he can select the second-best from the rest, etc. As a result, all alternatives appear to be ordered by preferences of an expert. Such an ordering is denoted as a ranking.

Let $A = \{a_1, \ldots a_N\}$ be some unordered set of alternatives. Let us enumerate alternatives just after the expert has ordered them by his preferences. Therefore, A is the ordered set represented by the strict $P = a_1 \succ a_2 \succ \ldots \succ a_N$ or non-strict $P = a_1 \succeq a_2 \succeq \ldots \succeq a_N$ ranking in general. It the last case, the expert cannot distinguish between some alternatives (he is not so strong in his preferences and has placed them as the same).

Therefore, the ranking means that for all pairs it can always be pointed what an alternative is better (not worse) than other. It is important to point out, the inverse proposition is incorrect in general case. The reconstruction of a ranking based on pairwise preferences is a special problem and is not investigated here.

The ranking P can be presented by the relation matrix $M_P(N, N)$ with elements

$$m_{ij} = \begin{cases} 1, a_i \succ a_j \\ 0, a_i \sim a_j \\ -1, a_i \prec a_j. \end{cases}$$

Let two rankings P_u and P_v be presented by relation matrices M_{P_u} and M_{P_v}. We can calculate distance between them under the assumption of the same elements numeration in A:

$$d(P_u, P_v) = \frac{1}{2} \sum_{i=1}^{N} \sum_{j=1}^{N} |m_{ij}^u - m_{ij}^v|. \tag{1}$$

It is known [1, 11], this value is a metric for binary relations of (quasi)orderings, i.e. for rankings.

Let n individual preferences (rankings) be given. It is needed to build a group relation P coordinated in a certain way with relations $P_1, \ldots P_n$. Methods for building the group relation are usually denoted as concordance principles. There are different concordance principles. If there is no limitations for concordance, then it doesn't matter what principle will be used.

The most popular principle is the majority rule [12]. Let rankings $P_1, \ldots P_n$ be given. Let in a pair of alternatives the preference $a_i \succ a_j$ be true. Then the value $n(i, j) = \sum_{u=1}^{n} (m = 1)_{ij}^u$ denotes the number of rankings, where experts have such a preference. One formulation of the majority rule can be presented in the form: if $n(i, j) \geqslant n(j, i)$, then $a_i \succ a_j$ for the group relation. For strict relations the majority rule is equivalent to $n(i, j) \geqslant n/2$. It is known in general case, the majority preference can be non-transitive even through all individual preferences are transitive.

The median P^* is the ranking with the least distance to other rankings

$$P^* = \arg\min_P \sum_{u=1}^{n} d(P, P_u). \tag{2}$$

It is proved, if the majority relation is transitive (or is transformed to this form by special procedures), then it appears to be the median, specifically, the Kemeny's median.

There are different algorithms to find the Kemeny's median. One version of the locally optimal algorithm to find the Kemeny's median [13] calculates the loss matrix $Q(N, N)$ for N alternatives.

Let some ranking P and expert's rankings $P_1, ...P_n$ be presented by relation matrices M_P and $M_{P_1}, ...M_{P_n}$. The distance from P to other rankings is defined by

$$\sum_{u=1}^{n} d(P, P_u) = \frac{1}{2} \sum_{u=1}^{n} \sum_{i=1}^{N} \sum_{j=1}^{N} |m_{ij} - m_{ij}^u| = \frac{1}{2} \sum_{i=1}^{N} \sum_{j=1}^{N} \sum_{u=1}^{n} d_{ij}(P, P_u),$$

where partial "distances" under condition $m_{ij} = 1$ are defined as

$$d_{ij}(P, P_u) = |m_{ij} - m_{ij}^u| = \begin{cases} 0, \ m_{ij}^u = 1, \\ 1, \ m_{ij}^u = 0, \\ 2, \ m_{ij}^u = -1. \end{cases} \tag{3}$$

A loss matrix element q_{ij} defines the total mismatch losses of the preference $a_i \succ a_j$ in the unknown ranking P relative to corresponding preferences in rankings $P_1, ...P_n$:

$$q_{ij} = \sum_{u=1}^{n} d_{ij}(P, P_u). \tag{4}$$

The Kemeny's algorithm finds the alternatives ordering to minimize the sum of elements above the main diagonal of the loss matrix Q.

1.2   The Problem of Immersing a Set in a Metric Space

Classical results about configurations of a set of points immersed in Euclidean space are well-known in multidimensional scaling theory. According to it, let the set $P_1, \ldots, P_n$ of rankings like points be immersed into Euclidean space as an unordered set.

This means the distance matrix $D(n, n)$ is calculated between all pairs of rankings, for example, according to (1). If metric violations haven't occurred in the set configuration, then it is known, the scalar product matrix calculated relative to some origin appears to be positive definite or nonnegative definite at least [14].

It doesn't matter, what point in the space can be used as the origin to calculate scalar products. It can be some element of the given set, or any other point elsewhere in the space. Nevertheless, according to the Torgerson's method of principal projections [15], it is suitable to use the center element of the given set as the origin. This center element P_0 can be presented by distances to other elements

$$d^2(P_0, P_i) = d_{0i}^2 = \frac{1}{n} \sum_{p=1}^{n} d_{ip}^2 - \frac{1}{2n^2} \sum_{p=1}^{n} \sum_{q=1}^{n} d_{pq}^2, \quad i = 1, \dots n. \tag{5}$$

The key problem of the multidimensional scaling theory is to restore the appropriate feature space of minimal dimensionality based on the distance matrix given. It is not our problem here.

Nevertheless, the expression (5) represents the arithmetic mean of a set of elements in the Euclidean space for data in the form of pairwise distances only (not features themselves). The formula (5) has been effectively used to develop clustering and machine learning algorithms based on distance or similarity matrices only [16].

The second summand in (5) is the set dispersion

$$\sigma^2 = \frac{1}{n} \sum_{i=1}^{n} d^2(P_0, P_i) = \frac{1}{n} \sum_{i=1}^{n} \left(\frac{1}{n} \sum_{p=1}^{n} d_{ip}^2 - \frac{1}{2n^2} \sum_{p=1}^{n} \sum_{q=1}^{n} d_{pq}^2 \right) = \frac{1}{2n^2} \sum_{p=1}^{n} \sum_{q=1}^{n} d_{pq}^2.$$

Unfortunately, partial distances (4) are some sort of artificial "distances". The specific using of (5) to represent rankings usually consists in impossibility to immerse correctly the set of rankings in a metric space. It must be pointed out, because of metric violations sometimes we cannot calculate distances (5). In general, the second summand in (5) can exceed the first one resulted in a complex value of a distance.

To prove that metric violations have not occurred or to correct distances in other case, we have developed methods of optimal corrections [8–10].

Specifically, according to them, we use the reduced formula (5) without the second summand in all cases, independently of metric violations. The reduced expression (5) represents the element denoted as P_{00} which is out of the given set convex cover. Therefore, the element P_{00} can be used as the new origin to calculate the scalar product matrix relative to it, to correct distances, and to find the center element P_0.

2   The Problem to Find a Metric Kemeny's Median

2.1   Correction of Partial Distances for Individual Expert's Rankings

The central element P_0 is the least distant from other elements in the set like the median P^* and needs to satisfy (2) as well as a ranking. Nevertheless, the median P^* is represented both by distances (1) to other rankings and by the ranking itself, while the central element P_0 is represented only by distances (5) and it doesn't exist as a ranking.

If these rankings coincide with each other $P^* = P_0$, then we can develop mathematically correct clustering and machine learning algorithms for ordering scales based on pairwise comparisons [16].

It is known, strict and non-strict rankings are measurements in ordering scales. The monotonic transformation is correct for this type of scales, and consists in alternatives reallocation on a numerical axis without changing their ordering relative to each other.

In general, the center P_0 and the ranking P^* are presented by their own distances to other elements $P_1, ... P_n$. Therefore, based on the appropriate monotonic transformation to prove $P^* = P_0$, we can show that these rankings are equivalent. We use the Kemeny's algorithm.

Let rankings P_u, P_0 and P^* be given with $\delta = d(P_0, P_u) - d(P^*, P_u) \neq 0$. We need to investigate different cases of δ.

1. Let $\delta > 0$. To remove (compensate) this difference, it is necessary to uniformly distribute the value of $\delta > 0$ between all nonzero elements of M_{P_u} and to calculate the new relation matrix with elements

$$m_{ij}^u = \begin{cases} +1 + 2\delta/k, & a_i \succ a_j, \\ 0, & a_i \sim a_j, \\ -1 - 2\delta/k, & a_i \prec a_j, \end{cases}$$

where $k = N^2 - N - N_0$ is the general number of nonzero elements without the main diagonal, N_0 is the number of zero non-diagonal elements $m_{ij}^u = 0$. It is evident, this new matrix doesn't change the expert's ranking P_u (we have no rights to change it). We use doubled value $2\delta/k$, since in (1) all differences are used twice. We can correct nonzero elements only, since each zero element indicates two alternatives are on the same place in the expert's ranking. As a result, the distance between modified median and other ranking increases and compensates $\delta > 0$.

In this case, for $m_{ij} = 1$ the preference $a_i \succ a_j$ in the unknown ranking P is penalized by the expert's ranking P_u to form partial distances

$$d_{ij}(P, P_u) = \begin{cases} 2\delta/k, & m_{ij}^u = +1 + 2\delta/k, \\ 1, & m_{ij}^u = 0, \\ 2 + 2\delta/k, & m_{ij}^u = -1 - 2\delta/k. \end{cases}$$

Let $\delta < 0$. To remove this difference, it is necessary to uniformly distribute the value $\delta < 0$ between nonzero elements of M_{P_u} too. Additionally, the set size of nonzero modified elements $k' = k - \Delta k$ is decreased. We find Δk coinciding elements $m_{ij}^u = m_{ij}^*$ in relation matrices of the expert's ranking and the median. Indeed, the partial distance has zero value $d_{ij}(P^*, P_u) = 0$ in this case. We cannot decrease it, since any negative value of m_{ij}^u obligatory increases the distance between P_u and P^*.

2. Let $-1 < 2\delta/k' < 0$. In this case to get the unchanged expert's ranking P_u, it is necessary to calculate the new relation matrix M_{P_u} with elements

$$m_{ij}^u = \begin{cases} +1 - |2\delta/k'|, \, a_i \succ a_j, \\ 0, \, a_i \sim a_j, \\ -1 + |2\delta/k'|, \, a_i \prec a_j, \\ m_{ij}^*, \, m_{ij}^u = m_{ij}^*. \end{cases}$$

Consequently, for $m_{ij} = 1$ the preference $a_i \succ a_j$ in the unknown ranking P is penalized by the expert's ranking P_u to form partial distances

$$d_{ij}(P, P_u) = \begin{cases} 0, \, m_{ij}^u = 1, \\ 1, \, m_{ij}^u = 0, \\ 2, \, m_{ij}^u = -1, \\ |2\delta/k'|, \, m_{ij}^u = +1 - |2\delta/k'|, \\ 2 - |2\delta/k'|, \, m_{ij}^u = -1 + |2\delta/k'|. \end{cases}$$

If the value $2\delta/k' \leq -1$, then, inevitably, the signs of some elements m_{ij}^u are inverted. Therefore, it is necessary to change the expert's ranking P_u. Since we cannot change an expert's opinion, it is necessary to change the median as the ranking P^*. Therefore, the ranking P_0 is different from the ranking P^*. As a result, we try to find another ranking P^* more suitable to P_0.

3. Let $-2 \leq 2\delta/k' \leq -1$. It is necessary to calculate new relation matrix M_{P^*} with elements

$$m_{ij}^* = \begin{cases} +1 - |2\delta/k'|, \, a_i \succ a_j, \\ 0, \, a_i \sim a_j, \\ -1 + |2\delta/k'|, \, a_i \prec a_j, \\ m_{ij}^u, \, m_{ij}^u = m_{ij}^*. \end{cases}$$

Consequently, for $m_{ij} = 1$ the preference $a_i \succ a_j$ in the unknown ranking P is penalized by the ranking P^* to form partial distances

$$d_{ij}(P, P^*) = \begin{cases} 0, \, m_{ij}^* = 1, \\ 1, \, m_{ij}^* = 0, \\ 2, \, m_{ij}^* = -1, \\ |2\delta/k'|, \, m_{ij}^* = +1 - |2\delta/k'|, \\ 2 - |2\delta/k'|, \, m_{ij}^* = -1 + |2\delta/k'|. \end{cases}$$

4. Let $2\delta/k' < -2$. Hence, $2\delta/k' + 2 < 0$. First, we distribute between k' nonzero elements as above the value -2 only. Hence, the new relation matrix M_{P^*} is calculated with elements

$$m_{ij}^* = \begin{cases} +1 - 2, \, a_i \succ a_j, \\ -1 + 2, \, a_i \prec a_j, \\ 0, \, a_i \sim a_j, \\ m_{ij}^u, \, m_{ij}^u = m_{ij}^*. \end{cases}$$

After that, all preferences for nonzero elements $m_{ij}^* = \pm 1$ have been inverted.

Next, we distribute the negative remainder $(2\delta/k' + 2)k' = 2\delta + 2k' < 0$ between the same k' nonzero elements in the relation matrix for the ranking P^*. Let the remainder for each nonzero element be $-1 < 2\delta/k' + 2 < 0$. In this case, we distribute this remainder and don't change the new ranking P^*. Hence, it is necessary to calculate the relation matrix M_{P*} with preferences $a_i \succ a_j$ inverted to $a_i \prec a_j$ and vice-versa for nonzero $m_{ij}^* = \pm 1$ elements

$$m_{ij}^* = \begin{cases} +1 - |2\delta/k' + 2|, \, a_i \succ a_j, \\ -1 + |2\delta/k' + 2|, \, a_i \prec a_j, \\ 0, \, a_i \sim a_j, \\ m_{ij}^u, \, m_{ij}^u = m_{ij}^*, \end{cases}$$

where for $m_{ij} = 1$ the preference $a_i \succ a_j$ in the unknown ranking P is penalized by the ranking P^* to form partial distances

$$d_{ij}(P, P^*) = \begin{cases} 0, \, m_{ij}^* = 1, \\ 1, \, m_{ij}^* = 0, \\ 2, \, m_{ij}^* = -1, \\ |2\delta/k' + 2|, \, m_{ij}^* = +1 - |2\delta/k' + 2|, \\ 2 - |2\delta/k' + 2|, \, m_{ij}^* = -1 + |2\delta/k' + 2|. \end{cases}$$

In all cases $2\delta/k' \le -1$ of correction of the relation matrix M_{P*} the ranking P^* has changed and appears to be another median to conform to the center P_0.

5. Let $2\delta/k' < -|\mathbf{C}|$, where $C \le -2$. This is the generalization of the case 4 above. Let $-|C| - 1 < 2\delta/k' < -|C|$. First, as mentioned above, the new relation matrix M_{P*} is calculated with elements

$$m_{ij}^* = \begin{cases} +1 - |C|, \, a_i \succ a_j, \\ -1 + |C|, \, a_i \prec a_j, \\ 0, \, a_i \sim a_j, \\ m_{ij}^u, \, m_{ij}^u = m_{ij}^*. \end{cases}$$

Next, it is necessary to calculate the relation matrix M_{P*} with preferences $a_i \succ a_j$ inverted to $a_i \prec a_j$ and vice-versa for nonzero $m_{ij}^* = \pm 1$ elements

$$m_{ij}^* = \begin{cases} -1 + |C| - |2\delta/k' + |C||, \, a_i \succ a_j, \\ +1 - |C| + |2\delta/k' + |C||, \, a_i \prec a_j, \\ 0, \, a_i \sim a_j, \\ m_{ij}^u, \, m_{ij}^u = m_{ij}^*, \end{cases}$$

where for $m_{ij} = 1$ the preference $a_i \succ a_j$ in the unknown ranking P is penalized by the ranking P^* to form partial distances

$$d_{ij}(P, P^*) = \begin{cases} 0, \, m_{ij}^* = 1, \\ 1, \, m_{ij}^* = 0, \\ 2, \, m_{ij}^* = -1, \\ \left|2 - |C| + |2\delta/k' + |C||\right|, \, a_i \succ a_j, \\ |C| - |2\delta/k' + |C||, \, a_i \prec a_j. \end{cases}$$

2.2   Calculation of a Loss Matrix

To calculate the loss matrix $Q(N, N)$ according to (4), it is necessary to use modified relation matrices for corresponding expert's rankings $P_1, ... P_n$ or modified relation matrices for ranking P^* instead of corresponding unchanged expert's relation matrices.

It is necessary to say, indistinguishable alternatives in the median relation can be the natural result of the loss matrix Q discrete essence. We can decrease their number in the metric Kemeny's median, since its loss matrix appears to be nondiscrete in general. We developed a procedure of Kemeny's median step-by-step transformations after making some corrections in individual expert's rankings to get the final metric one:

Step 0.

1. Calculate the central element P_0 and the Kemeny's median P^* for rankings P_i, $i = 1, ... n$.
2. Calculate distances $d(P^*, P_i)$, $i = 1, ... n$.
3. Calculate differences $\delta_i = d(P_0, P_i) - d(P^*, P_i)$, $i = 1, ... n$.
4. $u = 1$.

Step u. Correction of individual expert's rankings:

1. Modify the individual relation matrix M_{P_u} for difference δ_u. Individual relation matrices M_{P_i}, $i = 1, ... u$ have been modified, and M_{P_i}, $i = u + 1, ... n$ – not yet.
2. Calculate the modified loss matrix Q_u.
3. Calculate the median P_u^* and distances $d(P_u^*, P_i)$, $i = 1, ... n$.
4. Calculate differences $\delta_i = d(P_0, P_i) - d(P_u^*, P_i)$, $i = 1, ... n$.
5. Stop, if $u = n$, else $u = u + 1$.

Result. Denote $P_0^* = P_n^*$ as the metric Kemeny's median, where $P_0^* = P_0$.

3   Experiment and Discussion

3.1   Initial Data and a Kemeny's Median

Here we discuss a small example of real data to demonstrate some peculiarities in calculating of the metric Kemeny's median. The well-known expert organizations like Consumer Reports, J.D.Power, Auto Express, TUV Rheinland, etc., publish annual reviews and tests of varied production, such as cars. Usually, there are different partially intersected lists of different car models. We use here the limited set (Table 1) of premium-class cars presented simultaneously in 2015–16 years in different lists by the organizations mentioned above as experts (we use two digits for numbers of alternatives to distinguish them from positions in rankings).

Hence, car ratings in our short list are not absolute. For example, the first place isn't the same in original ratings because some other cars are not presented here, the second one here usually doesn't immediately follow the previous place in original ratings too, etc. At last, experts mentioned above are indexed in an arbitrary order. The Kemeny's median is calculated based on the loss matrix

$$
\begin{array}{r}
01 \\
02 \\
03 \\
04 \\
05 \\
06 \\
07 \\
08 \\
09 \\
10 \\
11 \\
12
\end{array}
\begin{pmatrix}
0 & 4 & 2 & 2 & 0 & 0 & 0 & 0 & 0 & 0 & 0 & 0 \\
4 & 0 & 2 & 2 & 4 & 2 & 4 & 0 & 2 & 0 & 0 & 0 \\
6 & 6 & 0 & 4 & 2 & 2 & 2 & 0 & 2 & 0 & 0 & 0 \\
6 & 6 & 4 & 0 & 4 & 2 & 4 & 0 & 2 & 0 & 0 & 0 \\
8 & 4 & 6 & 4 & 0 & 2 & 4 & 2 & 2 & 2 & 0 & 0 \\
8 & 6 & 6 & 6 & 6 & 0 & 4 & 2 & 2 & 0 & 0 & 2 \\
8 & 4 & 6 & 4 & 4 & 4 & 0 & 4 & 6 & 4 & 4 & 4 \\
8 & 8 & 8 & 8 & 6 & 6 & 4 & 0 & 2 & 4 & 2 & 0 \\
8 & 6 & 6 & 6 & 6 & 6 & 2 & 6 & 0 & 4 & 4 & 4 \\
8 & 8 & 8 & 8 & 6 & 8 & 4 & 4 & 4 & 0 & 4 & 2 \\
8 & 8 & 8 & 8 & 8 & 8 & 4 & 6 & 4 & 4 & 0 & 4 \\
8 & 8 & 8 & 8 & 8 & 6 & 4 & 8 & 4 & 6 & 4 & 0
\end{pmatrix}
$$

and is represented in Table 1 as the ranking P^*. The first group of indistinguishable alternatives 09 and 10 takes the same position 8.5 as the standard rank. Next, the second group of indistinguishable alternatives 07, 11 and 12 takes the same position 11 as the standard rank too.

3.2   Calculation of a Metric Kemeny's Median

Expert's rankings as points in the Euclidean space are organized in right configuration, since the matrix of pairwise distances between expert's rankings has no metric violations according to the technique, as stated above. Pairwise distances between expert's rankings with distances to the Kemeny's median and to the central element are presented in the matrix below. It is necessary to point out, we don't know the distance between P_0 and P^*:

$$
\begin{array}{r}
1 \\
2 \\
3 \\
4 \\
P^* \\
P_0
\end{array}
\begin{pmatrix}
0 & 20 & 46 & 68 & 22 & 29.228 \\
20 & 0 & 50 & 64 & 22 & 28.605 \\
46 & 50 & 0 & 34 & 28 & 22.633 \\
68 & 64 & 34 & 0 & 46 & 39.221 \\
22 & 22 & 28 & 46 & 0 & - \\
29.228 & 28.605 & 22.633 & 39.221 & - & 0
\end{pmatrix} . \tag{6}
$$

Differences δ between distances from each expert's ranking P_u to the central element P_0 and to the median P^* are presented in Table 2.

As we can see, all corrections don't change signs of nonzero elements in expert's relation matrices. Therefore, all experts' rankings are not changed to form the metric Kemeny's median. The metric Kemeny's median is calculated based on the loss matrix Q:

$$
\begin{array}{c}
01 \\ 02 \\ 03 \\ 04 \\ 05 \\ 06 \\ 07 \\ 08 \\ 09 \\ 10 \\ 11 \\ 12
\end{array}
\left(
\begin{array}{cccccccccccc}
0 & 4.21 & 2.21 & 2.21 & 0.21 & 0.21 & 0.21 & 0.21 & 0.21 & 0.21 & 0.21 & 0.21 \\
4.21 & 0 & 1.87 & 1.87 & 3.60 & 1.94 & 3.60 & 0.21 & 1.94 & 0.21 & 0.21 & 0.21 \\
6.21 & 6.55 & 0 & 4.21 & 1.94 & 1.94 & 1.94 & 0.21 & 1.94 & 0.21 & 0.21 & 0.21 \\
6.21 & 6.55 & 4.21 & 0 & 3.60 & 1.94 & 3.60 & 0.21 & 1.94 & 0.21 & 0.21 & 0.21 \\
8.21 & 4.82 & 6.48 & 4.82 & 0 & 2.21 & 3.60 & 2.21 & 1.94 & 2.21 & 0.21 & 0.21 \\
8.21 & 6.48 & 6.48 & 6.48 & 6.21 & 0 & 3.60 & 2.21 & 1.94 & 0.21 & 0.21 & 2.21 \\
8.21 & 4.82 & 6.48 & 4.82 & 4.82 & 4.82 & 0 & 4.82 & 6.55 & 4.82 & 4.82 & 4.82 \\
8.21 & 8.21 & 8.21 & 8.21 & 6.21 & 6.21 & 3.60 & 0 & 1.94 & 3.94 & 1.94 & 0.21 \\
8.21 & 6.48 & 6.48 & 6.48 & 6.48 & 6.48 & 1.87 & 6.48 & 0 & 4.82 & 4.22 & 4.21 \\
8.21 & 8.21 & 8.21 & 8.21 & 6.21 & 8.21 & 3.60 & 4.48 & 3.60 & 0 & 3.87 & 2.21 \\
8.21 & 8.21 & 8.21 & 8.21 & 8.21 & 8.21 & 3.60 & 6.48 & 4.21 & 4.55 & 0 & 4.82 \\
8.21 & 8.21 & 8.21 & 8.21 & 8.21 & 6.21 & 3.60 & 8.21 & 4.21 & 6.21 & 3.60 & 0
\end{array}
\right)
$$

and is denoted in Table 1 as the ranking P_0^*.

3.3 Discussion of Results

The metric Kemeny's median (denoted in Table 1 as P_0^*) appears without indistinguishable alternatives. As we remarked above, the loss matrix for the metric median usually consists of non-discrete values. In this case, the metric median doesn't change the classic median ranking P^*, but makes it more precise (see Table 3) formally resulting in a different binary relation as the ranking P_0^*. Therefore, the ranking P_0^* represents the central element $P_0^* = P_0$.

Table 1. Car ratings

Num	Model	Expert's rankings				P^*	P_0^*	P_1^*	P_2^*	P_3^*	P_4^*
01	Porsche	2	4	1	1	1	1	1	1	1	1
02	Lexus	1	1	6	6	2	2	2	2	2	2
03	Toyota	4	3	2	7	3	3	3	3	3	3
04	Buick	3	2	5	8	4	4	4	4	4	4
05	Chevrolet	6	7	4	4	5	5	5	5	5	5
06	Lincoln	8	5	7	5	6	6	6	6	6	6
07	BMW	12	12	3	3	11	12	11	11	12	12
08	GMC	5	8	8	11	7	7	7	7	7	7
09	Infiniti	11	11	9	2	8.5	9	8	8	9	9
10	Mercedes-Benz	10	6	11	9	8.5	8	9	9	8	8
11	Audi	9	10	10	10	11	11	11	11	11	11
12	Acura	7	9	12	12	11	10	11	11	10	10

Let us use the step-by-step transformations of the Kemeny's median P^* to show changes between two medians P^* and P_0^*. According to the procedure stated above, the Kemeny's median P_1^* to remove difference δ_1 relative to the first expert's ranking (Table 2) is calculated based on the loss matrix Q_1:

Table 2. Correction of relation matrices for expert's rankings

P^*	P_0	Difference (δ)	Number of modified elements (k)	Average $(2\delta/k)$
22	29.228	+7.228	132	+0.10951
22	28.605	+6.605	132	+0.10008
28	22.633	−5.367	32	−0.33544
46	39.221	−6.779	50	−0.27118

$$
\begin{array}{c}
01 \\ 02 \\ 03 \\ 04 \\ 05 \\ 06 \\ 07 \\ 08 \\ 09 \\ 10 \\ 11 \\ 12
\end{array}
\left(
\begin{array}{cccccccccccc}
0 & 4.11 & 2.11 & 2.11 & 0.11 & 0.11 & 0.11 & 0.11 & 0.11 & 0.11 & 0.11 & 0.11 \\
4.11 & 0 & 2.11 & 2.11 & 4.11 & 2.11 & 4.11 & 0.11 & 2.11 & 0.11 & 0.11 & 0.11 \\
6.11 & 6.11 & 0 & 4.11 & 2.11 & 2.11 & 2.11 & 0.11 & 2.11 & 0.11 & 0.11 & 0.11 \\
6.11 & 6.11 & 4.11 & 0 & 4.11 & 2.11 & 4.11 & 0.11 & 2.11 & 0.11 & 0.11 & 0.11 \\
8.11 & 4.11 & 6.11 & 4.11 & 0 & 2.11 & 4.11 & 2.11 & 2.11 & 2.11 & 0.11 & 0.11 \\
8.11 & 6.11 & 6.11 & 6.11 & 6.11 & 0 & 4.11 & 2.11 & 2.11 & 0.11 & 0.11 & 2.11 \\
8.11 & 4.11 & 6.11 & 4.11 & 4.11 & 4.11 & 0 & 4.11 & 6.11 & 4.11 & 4.11 & 4.11 \\
8.11 & 8.11 & 8.11 & 8.11 & 6.11 & 6.11 & 4.11 & 0 & 2.11 & 4.11 & 2.11 & 0.11 \\
8.11 & 6.11 & 6.11 & 6.11 & 6.11 & 6.11 & 2.11 & 6.11 & 0 & 4.11 & 4.11 & 4.11 \\
8.11 & 8.11 & 8.11 & 8.11 & 6.11 & 8.11 & 4.11 & 4.11 & 4.11 & 0 & 4.11 & 2.11 \\
8.11 & 8.11 & 8.11 & 8.11 & 8.11 & 8.11 & 4.11 & 6.11 & 4.11 & 4.11 & 0 & 4.11 \\
8.11 & 8.11 & 8.11 & 8.11 & 8.11 & 6.11 & 4.11 & 8.11 & 4.11 & 6.11 & 4.11 & 0
\end{array}
\right) .
$$

Other loss matrices Q_u to remove differences δ_u relative to other expert's rankings P_u are calculated in a similar way. At last, the final loss matrix Q_n is the loss matrix for the metric Kemeny's median $Q_n = Q$. All rankings are showed in Table 3.

Table 3. Car rankings

Type	1	2	3	4	5	6	7	8	9	10	11	12
P^*	01 ≻	02 ≻	03 ≻	04 ≻	05 ≻	06 ≻	08 ≻	09 ∼	10 ≻	07 ∼	11 ∼	12
P_0^*	01 ≻	02 ≻	03 ≻	04 ≻	05 ≻	06 ≻	08 ≻	10 ≻	09 ≻	12 ≻	11 ≻	07
P_1^*	01 ≻	02 ≻	03 ≻	04 ≻	05 ≻	06 ≻	08 ≻	09 ≻	10 ≻	07 ∼	11 ∼	12
P_2^*	01 ≻	02 ≻	03 ≻	04 ≻	05 ≻	06 ≻	08 ≻	09 ≻	10 ≻	07 ∼	11 ∼	12
P_3^*	01 ≻	02 ≻	03 ≻	04 ≻	05 ≻	06 ≻	08 ≻	10 ≻	09 ≻	12 ≻	11 ≻	07
P_4^*	01 ≻	02 ≻	03 ≻	04 ≻	05 ≻	06 ≻	08 ≻	10 ≻	09 ≻	12 ≻	11 ≻	07
P_1	02 ≻	01 ≻	04 ≻	03 ≻	08 ≻	05 ≻	12 ≻	06 ≻	11 ≻	10 ≻	09 ≻	07
P_2	02 ≻	04 ≻	03 ≻	01 ≻	06 ≻	10 ≻	05 ≻	08 ≻	12 ≻	11 ≻	09 ≻	07
P_3	01 ≻	03 ≻	07 ≻	05 ≻	04 ≻	02 ≻	06 ≻	08 ≻	09 ≻	11 ≻	10 ≻	12
P_4	01 ≻	09 ≻	07 ≻	05 ≻	06 ≻	02 ≻	03 ≻	04 ≻	10 ≻	11 ≻	08 ≻	12

We can see (Tables 1, 3) with the first expert's modified relation matrix alternatives 09 and 10 are distinguished by the median 09 ≻ 10, while alternatives

07, 11 and 12 are not. With the first and the second expert's modified relation matrices the median produces the same ranking. At last, with the first three and all expert's modified relation matrices the median produces all distinguished alternatives with reversed relation $10 \succ 09$.

Therefore, all rankings representing variants of the metric Kemeny's median with the final ranking don't change the ranking of the classic Kemeny's median, but make the median preferences more precise step-by-step. Such rankings are similar, but formally appear to be different binary relations.

In general, we can show both Kemeny's medians are the same, if there are no indistinguishable alternatives in the classic Kemeny's median.

Let us investigate more closely expert's rankings P_i, $i = 1, \ldots 4$ (Table 3). As we can see, expert's rankings can be split on two groups: P_1, P_2 and P_3, P_4. The first group represents preference $10 \succ 9$ with the alternative 07 on the last position. The second group represents the opposite preference $9 \succ 10$ with the alternative 07 at the beginning of ranking on the third position. As a result, according to the distance matrix (6), rankings in groups are less distant from each other than rankings from different groups.

It is known, the Kemeny's median for two opposite rankings consists in all indistinguishable alternatives in this specific case. Such an "ordering" as a binary relation is placed exactly in the middle between two opposite relations. In general, appearance of indistinguishable alternatives in the Kemeny's median for some set of expert's rankings is the basis for the idea of different experts' groups (clusters) appearance with generally different preferences.

Therefore, some indistinguishable alternatives in the Kemeny's median are the result of mutual compensations of partially opposite preferences by discrete (integer) partial distances (3) in the loss matrix Q. As a result, the more size of the experts' set is, the more actual the clustering problem will be.

The idea of the metric Kemeny's median is the correct basis for the following technique. We use, for example, the k-means algorithm in the specially developed mathematically correct form for a set of elements represented by mutual distances only [16] to cluster experts' rankings. As a result, we restore corresponding rankings for cluster centers. Using this technique, we can investigate rankings in terms of standard clustering problem.

4   Conclusion

The arithmetic mean as the center of the given rankings set is not presented as a proper ranking. In general, the central element differs by its distances from the corresponding distances of the classic Kemeny's median to other elements. We find the metric Kemeny's median, which coincides with the arithmetic mean of the given set by distances to other elements.

As a ranking, the metric Kemeny's median coincides with the classic one, if the classic case is mathematically correct. The metric Kemeny's median is the correct center of the rankings set.

According to the idea of correct center, the proposed approach realizes the mathematically correct technique to investigate the decision-making problem (rank aggregation, group ranking, expert's concordance, etc.) as the usual k-means clustering problem for ordering scales. We simply use the k-means in the appropriate realization [16] for the metrically correct matrix of pairwise comparisons and restore object orderings for cluster centers at the end of this process.

The complexity of this procedure consists in the known complexity of k-means and Kemeny consensus problems.

Today, clustering becomes actual demand for the Kemeny's median in the field of big data processing because of many human experts and AI expert systems supporting, for example, relevant medical diagnostics, etc.

Future work consists in investigating more closely vicinities of metric medians and boundaries between clusters for data in ordering scales.

Acknowledgments. This research was partially supported by the Russian Foundation for Basic Research (RFBR) grants 15-05-02228, 15-07-08967, 17-07-00319, 17-07-00436.

References

1. Litvak, B.G.: Expert Information: Methods of Acquisition and Analysis, 184 p. Radio i svyaz, Moscow (1982). (in Russian)
2. Charon, I., Guenoche, A., Hudry, O., Woirgard, F.: New results on the computation of median orders. Discrete Math. **165**(166), 139–153 (1997). https://doi.org/10.1016/S0012-365X(96)00166-5
3. Biedl, T., Brandenburg, F.J., Deng, X.: Crossings and permutations. In: Healy, P., Nikolov, N.S. (eds.) GD 2005. LNCS, vol. 3843, pp. 1–12. Springer, Heidelberg (2006). https://doi.org/10.1007/11618058_1
4. Conitzer, V., Davenport, A., Kalagnanam, J.: Improved bounds for computing Kemeny rankings. In: Proceedings of the 21st National Conference on Artificial Intelligence, vol. 1, pp. 620–626 (2006). http://www.cs.cmu.edu/conitzer/kemenyAAAI06.pdf
5. Nogin, V.D.: Reducing of Pareto Set: an Axiomatic Approach. Fizmatlit, Moscow (2016). (in Russian)
6. Larichev, O.I., Moshkovich, E.M.: Qualitative Methods of Decision Making. Verbal Analysis of Decisions. Nauka, Fizmatlit, Moscow (1996). (in Russian)
7. Jiao, Y., Korba, A., Sibony, E.: Controlling the distance to a Kemeny consensus without computing it. In: Balcan, M.F., Weinberger, K.Q. (eds.) Proceedings of The 33rd International Conference on Machine Learning. PMLR, vol. 48, pp. 2971–2980 (2016)
8. Dvoenko, S.D., Pshenichny, D.O.: Optimal correction of metrical violations in matrices of pairwise comparisons. JMLDA **1**(7), 885–890 (2014). (in Russian)
9. Dvoenko, S.D., Pshenichny, D.O.: On metric correction of matrices of pairwise comparisons. JMLDA **1**(5), 606–620 (2013). (in Russian)
10. Dvoenko, S.D., Pshenichny, D.O.: A recovering of violated metrics in machine learning. In: Proceedings of the Seventh Symposium on Information and Communication Technology (SoICT'16), pp. 15–21. ACM, New York (2016). https://doi.org/10.1145/3011077.3011084

11. Kemeny, J., Snell, J.: Mathematical Models in the Social Sciences. Blaisdell, New York (1963)
12. Mirkin, B.G.: The Problem of a Group Choice. Nauka, Moscow (1974). (in Russian)
13. Kemeny, J.: Mathematics without numbers. Daedalus **88**(4), 577–591 (1959)
14. Young, G., Housholder, A.S.: Discussion of a set of points in terms of their mutual distances. Psychometrica **3**(1), 19–22 (1938). https://doi.org/10.1007/BF02287916
15. Torgerson, W.S.: Theory and Methods of Scaling, 460 p. Wiley, New York (1958). https://doi.org/10.1002/bs.3830040308
16. Dvoenko, S.D.: Clustering and separating of a set of members in terms of mutual distances and similarities. Trans. MLDM **2**(2), 80–99 (2009)

Intelligent Data Processing in Life and Social Sciences

Recognition of Herpes Viruses on the Basis of a New Metric for Protein Sequences

Valentina Sulimova[1]([✉]), Oleg Seredin[1], and Vadim Mottl[2]

[1] Tula State University, Tula, Russia
vsulimova@yandex.ru, oseredin@yandex.ru
[2] Federal Research Center "Computer Science and Control" of RAS, Moscow, Russia
vmottl@yandex.ru

Abstract. This paper addresses the problem of intellectual human herpes viruses recognition based on the analysis of their protein sequences. To compare proteins, we use a new dissimilarity measure based on finding an optimal sequence alignment. In the previous work, we proved that the proposed way of sequence comparison generates a measure that has properties of a metric. These properties allow for more convenient and effective use of the proposed measure in further analysis in contrast to the traditional similarity measure, such as Needleman-Wunch alignment. The results of herpes viruses recognition show, that the metric properties allow to improve the classification quality. In addition, in this paper, we adduce an updated computational scheme for the proposed metric, which allows to speed up the comparison of proteins.

Keywords: Metric for protein sequences · Optimal sequence alignment · Two-class recognition · SVM · Herpes viruses

1 Introduction

Human herpes viruses are pathogens that establish lytic and latent infections. These viruses usually are not life-threatening, but in some cases they might cause serious infections of eyes and brain that can lead to blindness and, possibly, death. An effective drug (acyclovir and its derivatives) is available against these viruses. Therefore, early detection and identification of these viral infections is highly important for an effective treatment [1,2].

In this work, we propose the way to recognize herpes viruses basing only on their protein sequences.

It is evident that in this case the sequence comparison quality plays an important role for the resulting quality of recognition.

Traditionally, comparison of proteins consists in computing similarity measures based on finding an optimal pair-wise alignment [3–6]. But they don't

© Springer Nature Switzerland AG 2019
V. V. Strijov et al. (Eds.): IDP 2016, CCIS 794, pp. 61–73, 2019.
https://doi.org/10.1007/978-3-030-35400-8_5

provide the possibility for using advantages of popular and effective linear methods initially designed for feature spaces, like SVM for the two-class pattern recognition problem [7].

Partially, the problem of harnessing linear methods in featureless situations can be solved by constructing special similarity measures called positive definite kernel functions [8–10], which embed a set of proteins into some hypothetical linear space and play the role of inner product in it [10]. But constructing mathematically correct and at the same time biologically motivated kernel functions is, as a rule, theoretically and computationally hard problem [10–12]. Besides, there are entire continuous classes of kernels, which are equivalent from the viewpoint of the final decision rule [13].

At the same time, obviously, it is not the feature vectors of entities in some linear space that are the actual basis of machine learning and data mining algorithms but rather the respective metric, i.e., pairwise distance between entities [14]. In this connection there is a natural desire to use comparison measures possessing metric properties, especially since a metric allows for embedding any set of proteins into a linear space, and apply linear methods in it in accordance with the generalized linear approach to dependence estimation [15].

However, the dependence-estimation methods effectiveness crucially depends on the choice of the metric between entities, which has to satisfy the compactness hypothesis [16,17]. This means that the values of the accepted metric between the proteins performing the same function in a living organism must be small and, respectively, they must be large for proteins that perform different functions.

We know a number of ways to introduce metrics on a set of sequences [14], but in the case of amino acid sequences none of them has satisfactory interpretation from the biological point of view. In this connection, it is unlikely that the compactness hypothesis would hold true in the respective metric space of proteins. This point has multiple confirmations in practice [18,19], which gave rise to an entire series of papers aimed at improving the initial metric via various algebraic constructions of different complexity rate (Metric Learning), [17–21], including constructing metrics on the basis of kernel functions (Metric Kernel Learning) and constructing secondary features or new metrics on the basis of some similarity measures. Also, a number of papers tries to involve structural and protein-protein interaction information [22,23]. However, in this paper our goal is to provide recognition of human herpes viruses via constructing appropriate metrics for protein sequences without involving any additional information.

In this paper, we describe a simple way of constructing a mathematically correct metric on proteins set. This method follows the traditional Needleman-Wunsch algorithm, it is based on finding a global optimal alignment of sequences, and leans upon the probabilistic model PAM (Point Accepted Mutation) [24] of single amino acids. But, at the same time, it differs from the traditional approaches in the optimization criterion and in the way of comparing amino acids. In the previous paper [26], we proved that the proposed comparison measure possesses the properties of a metric. These properties allow for more convenient and effective use of the initial comparison measure.

In this paper, we adduce an updated computational scheme for the proposed metric, which allows to speed up the process of protein comparison. It is essential for real data processing.

This paper contains a more detailed experimental study of the proposed metric. Our results have shown that the metric's properties allow for improving the recognition quality of herpes viruses as distinct from the traditional Needleman-Wunsch procedure.

2 Comparison of Protein Sequences

2.1 A Metric on the Set of Protein Sequences

Let Ω be the set of all protein sequences. In this work, we consider only the primary structures of proteins, thus, we consider sequences over the alphabet of 20 known amino acids $A = \{\alpha^1, \ldots, \alpha^m\}, m = 20$.

Let also $\omega' = (\alpha'_1, \alpha'_2, \ldots, \alpha'_{N'}) \in \Omega$ and $\omega'' = (\alpha''_1, \alpha''_2, \ldots, \alpha''_{N''}) \in \Omega$ be two sequences of lengths N' and N'' that contain amino acids $\alpha'_i, \alpha''_j \in A$, $i = 1, \ldots, N'$, $j = 1, \ldots, N''$.

It is evident that any protein sequences comparison should be based on the comparison of amino acids forming them. The main theoretical concept underlying the proposed amino acids comparison method is the well-known probabilistic model of amino acids evolution by Margaret Dayhoff, called PAM (Pointed Accepted Mutation) [24]. Its main instrument is the notion of the Markov chain of amino acids evolution at some single point of protein's chain, which is represented by the matrix of transitional probabilities $\Psi = (\psi_{[1]}(\alpha^j|\alpha^i))$ of changing amino acid α^i into amino acid α^j at the next step of evolution. The index "1" in the brackets means that the initial one-step Markov chain is considered.

In accordance with the PAM model, it is supposed that this Markov chain is an ergodic and reversible random process, i.e., a process that possesses a final probability distribution

$$\xi\left(\alpha^j\right) = \sum_{\alpha^i \in A} \xi\left(\alpha^i\right)\psi_{[1]}\left(\alpha^j|\alpha^i\right)$$

and satisfies the reversibility condition

$$\xi(\alpha^i)\psi_{[1]}(\alpha^j|\alpha^i) = \xi(\alpha^j)\psi_{[1]}(\alpha^i|\alpha^j).$$

Let us consider the probabilistic process of evolution with a greater evolutionary step $s > 1$, i.e., a sparse Markov chain with the matrix of transitional probabilities $\Psi_{[s]} = [\underbrace{\Psi_{[1]} \times \ldots \times \Psi_{[1]}}_{s}]$. Previously we proved [10] that, for any s, similarity measures $\kappa_s(\alpha^i, \alpha^j) = \psi_{[s]}(\alpha^i|\alpha^j)/\xi(\alpha^i)$ form nonnegative definite matrixes of amino acids pairwise similarity. So, each of them is kernel function that embeds the set of amino acids A into a respective hypothetical linear space $\tilde{A}_s \subset A$ with Euclidean metric [25]

$$\rho_s(\alpha^i, \alpha^j) = \left(\kappa_s(\alpha^i, \alpha^i) + \kappa_s(\alpha^j, \alpha^j) - 2\kappa_s(\alpha^i, \alpha^j)\right)^{1/2}, \quad s = 1, 2, \ldots. \tag{1}$$

This is the same metric we use here to compare amino acids. Hereafter, we shall omit lower index $\rho(\alpha^i, \alpha^j)$, $\kappa(\alpha^i, \alpha^j)$, assuming that the evolutionary step $s > 1$ is predefined.

Further, we define a metric on the amino acid sequences set on the basis of global pairwise sequence alignment $\mathbf{w}(\omega', \omega'')$, which is understood as a way of sequence transformation by inserting gaps in some positions of aligned sequences for reducing them to a common length with preserving correspondences between their elements.

If the position $\mathbf{w}_i, i = 1, \ldots, |\mathbf{w}|$, of an alignment $\mathbf{w}(\omega', \omega'')$ of two sequences $\omega' = (\alpha'_1, \ldots, \alpha'_{N'}) \in \Omega$ and $\omega'' = (\alpha''_1, \ldots, \alpha''_{N''}) \in \Omega$ does not contain a gap, it explicitly defines two aligned amino acids $(\alpha'_{\mathbf{w}_{i,1}}, \alpha''_{\mathbf{w}_{i,2}})$.

We define the comparison measure of any two sequences as the optimal value of the specific optimization criterion

$$r(\omega', \omega'') = \min_{\mathbf{w}} \sqrt{\sum_{i=1}^{|\mathbf{w}|} \left[I(\mathbf{w}_i)\beta^2 + (1 - I(\mathbf{w}_i))\,\rho^2(\alpha'_{\mathbf{w}_{i,1}}, \alpha''_{\mathbf{w}_{i,2}}) \right]}, \qquad (2)$$

where $I(\mathbf{w}_i) = 1$ if the i-th position of the alignment $\mathbf{w}$ contains a gap and $I(\mathbf{w}_i) = 0$ otherwise, and coefficient β in (2) has the meaning of gap penalty. In our previous work [26] we proved that for any

$$\beta \geq 0.5 \max_{\alpha', \alpha'' \in A} \rho(\alpha', \alpha''), \quad \forall \alpha \in A$$

the comparison measure (2) possesses all the properties of a metric.

2.2 Dynamic Programming Procedure for Computing the Metric on the Set of Protein Sequences

The criterion (2) falls into the class of pairwise separable goal functions. A minimum of such function can be found using dynamic programming procedure, which is similar to the Needleman-Wunsch procedure [3] for finding the optimal global alignment of any two amino acid sequences, which maximizes their similarity.

The idea of this algorithm consists in recurrent computing of unknown dissimilarity values $F_{i,j}$ for growing fragments of two protein sequences $(\alpha'_1, \alpha'_2, \ldots, \alpha'_i)$ and $(\alpha''_1, \alpha''_2, \ldots, \alpha''_j)$ on the basis of dissimilarity values, which are assumed to be already computed:

$$F_{i,j} = \min \begin{cases} F_{i-1,j-1} + \rho^2(\alpha'_i, \alpha''_j); \\ F_{i-1,j} + \beta^2; \\ F_{i,j-1} + \beta^2, \end{cases} \quad i = 1, \ldots, N', \quad j = 1, \ldots, N''.$$

The computation starts from the initialization:

$$F_{0,0} = 0; \quad F_{i,0} = i\beta^2, \ i = 1, \ldots, N'; \quad F_{0,j} = j\beta^2, \ j = 1, \ldots, N'',$$

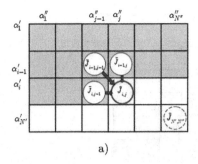
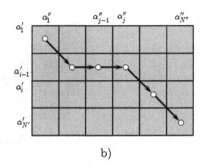

a)

b)

Fig. 1. A table of pair-wise correspondences between elements of two sequences being compared: (a) an illustration of the computation process, (b) a possible optimal alignment

and is finished, when the end of sequences is met: $r(\omega', \omega'') = \sqrt{F_{N'N''}}$. It is convenient to represent such a computation process as a table of pairwise correspondences (Fig. 1).

The computation process consists in consecutive passing through all the cells of the table (Fig. 1a), from the top left cell to the right bottom one, making recurrent computations of incomplete dissimilarity values $F_{i,j}$, choosing (and possibly saving) the direction of the optimal movement to the current cell (horizontal, vertical or diagonal), which are to be used later to find the optimal alignment (Fig. 1b).

2.3 Speeding-Up the Metric Computation for Two Protein Sequences

It is evident that the computational complexity of the dynamic programming procedure described in the previous section is proportional to the product $N'N''$. So, the computing time is essentially increasing, when the lengths of sequences increase. As a result, there is the problem of applying the described algorithm to real data, which are usually represented by big sets of long protein sequences. This is a well-known problem, and there is a number of heuristic realizations of similar dynamic programming procedures for comparing biological sequences (such as BLAST, FASTA, and etc.) [27–29], that allow to obtain a fast but approximate solution to the dynamic programming problem.

To speed-up the comparison of protein sequences, we adapt here the approach that was proposed for comparing discrete signals with the purpose of speech recognition [30,31].

In accordance with this approach, we find the optimal alignment among only those alignments, which fully fall into a pruned $2t$-width subtable of pairwise alignments (Table 2).

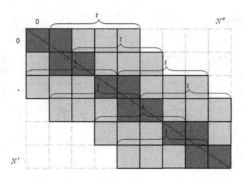

Fig. 2. A pruned $2t$-width table of pair-wise alignments

The respective modified recurrent computation scheme can be expressed in the form

$$F_{0,0} = 0; \quad F_{i,0} = i\beta^2, \ i = 1, \dots, N'; \quad F_{0,j} = j\beta^2, \ j = 1, \dots, \lfloor \frac{N''}{t} \rfloor,$$

$$F_{i,j} = \min \begin{cases} F_{i-1,j-1} + \rho^2(\alpha'_i, \alpha''_j); \\ F_{i-1,j} + \beta^2; \\ F_{i,j-1} + \beta^2, \end{cases} \quad i = 1, \dots, N', \quad j = j_{st}(i), \dots, j_{end}(i),$$

where

$$j_{st}(i) = \max \left(1, \lfloor \frac{N''}{N'}i - \frac{N''}{t} \rfloor \right), \quad j_{end}(i) = \min \left(N'', \lfloor \frac{N''}{N'}i + \frac{N''}{t} \rfloor \right).$$

It should be noticed that, in the strict sense, the proposed approach does not guarantee finding the optimal value of the criterion (2). But, at the same time, the strip width parameter t allows to control the number of alignments under consideration and, thus, to balance the accuracy and the speed of computation. Our experiments with real protein sequences of Herpes Viruses, described in Sect. 3 of this paper, have shown that the initial algorithm and its approximations with $t = 1, \dots, 1.5$ have absolutely the same results for all the pairs of 1532 considered proteins, the lengths of which range from 36 to 1378 amino acids.

Filling up the pruned table is simple enough, which results in increasing the computation speed even for essentially different sequences, in contrast to a more intellectual approach that was proposed for signals [32].

2.4 Speeding-Up the Computation of a Metric Values Matrix for Proteins Set

Let $\Omega = \{\omega_1, \dots, \omega_K\}$ be some set of K protein sequences to be analyzed. It should be noticed, that for many problems of protein analysis including herpes virus recognition, it is needed to compute the entire metric values matrix for the given set of proteins. However, due to the metric's symmetry, it is enough to compute the triangular matrix of metric values.

Moreover, it is easy to see that all the elements of such a matrix could be computed independently. Following the latter remark, we involve the technology of parallel computing OpenMP for further computation speeding-up. We have used a simple enough scheme of parallel computing, according to which for each row containing more than h elements, p threads have been created for parallel computing elements of the respective row.

3   Experiments

3.1   Data Description

For the experiments, we have used six sets of herpes virus amino acid sequences that perform functions specified in Table 1. These are the sets from the VIDA database (Virus Database at University College London) [33]. Proteins of each set have been partitioned into classes and homologous protein families on the basis of laboratory research of herpes viruses [2].

Besides, for some experiments we used an additional set of 143 protein sequences, which are not herpes viruses. These proteins were randomly chosen from the data set collected by Lanckriet et al. [34].

The resulting data set contains 1532 protein sequences of essentially different length in the range from 36 to 1378 amino acids.

Table 1. Experiment data structure

Set (num. of proteins)	Function	Class (num. of proteins)	Description	Homologous Protein Families (HPFs)
1 (233)	Membrane/ glycoprotein	1 (109)	Glycoprotein H	12, 42, 531
		2 (76)	Glycoprotein L	47, 50, 114, 256
		3 (48)	Glycoprotein M	20
2 (407)	Nucleotide/ repair metabolism	1 (256)	Thymidine kinase	2, 27
		2 (83)	Alkaline exonuclease	11, 51
		3 (37)	Ribonucleotide reductase	33
		4 (31)	dUTPase	43
3 (262)	Virion Assembly	1 (54)	Transport/capsid assembly protein	7
		2 (92)	DNA packaging protein	18, 22
		3 (77)	Cleavage/packaging protein	34, 39
		4 (20)	Packaging and capsid formation	79
		5 (19)	DNA pack. and capsid formation	108
4 (99)	Enzyme	1 (89)	Protein kinase	29, 40
		2 (10)	Phospholipase-like protein	328, 329

(contniued)

Table 1. *(contniued)*

Set (num. of proteins)	Function	Class (num. of proteins)	Description	Homologous Protein Families (HPFs)
5 (144)	DNA replication	1 (52)	Origin binding protein	5, 152
		2 (22)	DNA polymerase processivity factor	104, 1003
		3 (48)	Helicase/primase associated protein	16
		4 (22)	Component of DNA helicase/ promasecomplex	72
6 (195)	Virion protein	1 (47)	Virion tegument protein	21
		2 (28)	Tegument protein/FGARAT	44
		3 (21)	Tegument phosphoprotein	65
		4 (91)	Tegument protein	83, 86, 87, 93
		5 (29)	Virion protein	62, 106

3.2 Experimental Investigation of Metric Computing Efficiency

For experimental investigation of metric computing efficiency, we compared the initial dynamic programming procedure (Sect. 2.2) and its approximation with different values of the width parameter t (Sect. 2.3), where $t = 1$ corresponds to the initial full-width procedure. Additionally, each of these procedures was tested with 1, 2, and 4 threads on the personal computer of the following configuration: Intel Core i5-4210U CPU 1.70 GHz, 2 processor cores with the Hyper Threading, 6 Gb RAM, 36 Mb L3 cache.

The time for computing the whole matrix of the metric values for 1532 protein sequences was measured in minutes 5 times for each set of parameters. The average results are presented in Table 2.

Table 2. Average time (in minutes) for computing the matrix of the metric for 1532 proteins for different computation schemes and parameters of the computation

Number of threads (p)	$t = 1$	$t = 1.2$	$t = 1.5$	$t = 2$
$p = 1$	82.01	74.64	65.57	53.44
$p = 2$	42.15	39.27	35.43	30.61
$p = 4$	34.61	31.23	28.74	23.29

It is evident that the speed increases when the value t is increasing. Nevertheless, we have found that the initial algorithm and its heuristic approximations with $t = 1, \ldots, 1.5$ showed absolutely the same results for all the pairs of 1532 proteins studied in the experiment.

So, the proposed modifications of the computation scheme allow for decreasing the computation time from 82.01 to 28.74 min. without accuracy loss (for $t = 1.5$ and $p = 4$ threads). The acceleration in this case is about 2.85. This result is good enough for the architecture with only 2 real processor cores.

3.3   Investigation of Metric Properties Usability for Herpes Viruses Recognition

Constructing Kernel Functions on the Basis of Comparison Measures.
The experiment's main purpose is to demonstrate that the presence of comparison measure metric properties allows to increase the recognition quality. So, for this purpose, we have compared two very similar comparison measures, each of them is based on the global optimal pair-wise alignment of sequences - the Needleman-Wunch (NW) alignment and the proposed metric. These two comparison measures have similar structures, but while the first is a similarity measure, the second one possesses the metric properties.

In our experiments, we used NW algorithm from the MATLAB bioinformatics toolbox with both PAM250 and BLOSSUM62 substitution matrixes for comparing single amino acids and with the gap penalties g = 12 for start and b = 1 for continuation of gaps series.

The metric on the set of amino acids was constructed in accordance with (1) on the basis of PAM model with the evolutionary step $s = 250$. The parameter β of the proposed metric was set to $\beta = 0.0234$ for all the experiments.

The proposed metric, as it turned out, is the Euclidean metric for the considered sets of proteins. This fact gave us the possibility to apply the radial kernel function $K_{mtr}^{\alpha} = K_{mtr}^{\alpha}(\omega', \omega'') = \exp(-\alpha r^2(\omega', \omega''))$ [9,23]. But it should be noticed, that, in general case, the proposed way of metric construction doesn't guarantee the Euclidean property of the proposed metric and, thus, such transformation can lead to the presence of negative eigenvalues of the respective kernel matrix for proteins set. But the practice shows that the Euclidean property usually holds true.

Also we have constructed linear kernel functions in the space of the respective secondary features $K_{mtr}^{SF} = R_{mtr}^{T} R_{mtr}$, where $R_{mtr} = \{r(\omega_i, \omega_j), i, j = 1, \ldots, 1532\}$ is the matrix of metric values.

As to the NW similarity measure, the respective radial function was not used for it at all, because it is a similarity measure, and, moreover, it often yields negative values. So, any heuristic transformation of such similarity measure into a metric requires moving its values into the positive range. This fact requires having a full set of proteins at the training stage, but usually it is impossible in practice. In this connection, for the NW similarity measure we constructed linear kernel functions in the space of the respective secondary features $K_{NW}^{SF} = S_{NW}^{T} S_{NW}$, where S_{NW} is a NW protein similarity matrix.

Recognition of Classes and HPFs of Herpes Viruses. For this experiment we used two sets of Herpes viruses amino acid sequences that perform, respectively, the following functions: "Membrane/glycoprotein" and "Nucleotide repair/metabolism". It is important to distinguish these classes and HPFs from one another, the more so because previous investigations showed them to be most problematic ones for recognition [2].

For each indicated way of proteins comparison and for each of the two considered proteins sets, a number of two-class pattern recognition problems were solved (one-against-all and one-against-one recognition for classes and for HPFs). In each case, the training was made via SVM [7]. The quality of obtained decision rules was estimated by the leave-one-out cross validation (LOO) procedure. Table 3 contains LOO-error percentages only for the cases with results that differ at least for two algorithms. Besides, the results obtained for NW with PAM250 and BLOSSUM62 are practically the same in these experiments. In this connection, we included only one row for NW algorithm, which corresponds to the BLOSSUM62 substitution matrix.

Table 3. LOO percentages for recognition of the most interesting HPFs and classes

Membrane/glycoproteins											
Comparing measure	One-against-all								One-against-one		
	HPFs					Classes			HPFs		
	12	20	47	50	114	1	2	3	531/32	531/47	531/12
K_{NW}^{SF}	15.02	0.43	4.72	0.43	4.72	0.86	0.86	0.43	6.45	4.17	48.6
K_{mtr}^{SF}	15.40	0.00	0.00	0.00	0.43	0.43	0.43	0.00	3.22	2.08	50.0
$K_{mtr}^{0.01}$	14.59	0.00	0.00	0.00	0.43	0.43	0.43	0.00	3.22	2.08	50.0
Nucleotide repair/metabolism											
Comparing measure	One-against-all					One-against-one					
	HPF	Classes				HPFs		Classes			
	15	1	2			51/2	51/27	1/2	2/3		
K_{NW}^{SF}	14.25	0.49	0.49			1.91	1.91	0.59	1.67		
K_{mtr}^{SF}	14.49	0.49	0.25			0.64	0.64	0.3	1.67		
$K_{mtr}^{0.01}$	14.09	0.25	0.25			0.64	0.64	0.3	0.83		

Recognition of Herpes Viruses Proteins Performing a Specified Function. In this experiment, we solved 12 two-class protein recognition problems: tasks 1–6 for recognition each of 6 proteins sets (Table 1) from not Herpes viruses, and tasks 7–12 for recognition each of 6 sets from all other proteins. The classical SVM was used for constructing decision rules. The quality of decisions was estimated by ROC-scores computed for 10-times cross-validation with random forming training and testing sets in the proportion 20:80. The results are presented in Table 4. The best result for each task is shown in bold.

Table 4. Average ROC scores for 12 tasks of recognition Herpes viruses

Comp. measure	Tasks											
	1	2	3	4	5	6	7	8	9	10	11	12
K_{NW}^{SF}	0.923	0.823	0.758	0.932	0.804	0.856	0.856	0.973	0.935	0.915	0.938	0.927
K_{mtr}^{SF}	0.909	0.848	0.747	0.972	0.790	0.856	0.874	0.968	0.978	0.882	0.950	0.933
$K_{mtr}^{0.5}$	0.926	0.926	0.883	0.966	0.894	0.887	0.960	0.970	0.999	**0.997**	0.990	0.991
$K_{mtr}^{0.2}$	0.962	0.942	**0.907**	**0.985**	**0.927**	0.929	**0.966**	0.973	**1.000**	**0.997**	**0.994**	**0.996**
$K_{mtr}^{0.1}$	**0.966**	**0.948**	0.898	0.981	0.917	**0.937**	0.959	0.970	0.998	0.996	0.993	0.993
$K_{mtr}^{0.01}$	0.965	0.944	0.870	0.975	0.826	0.935	0.943	0.969	0.995	0.994	0.987	0.986
$K_{mtr}^{0.001}$	0.965	0.943	0.868	0.974	0.824	0.935	0.942	0.970	0.995	0.994	0.987	0.985

The obtained results have shown that the use of proposed metric instead of the traditional NW-similarity measure allows for essential increasing in the Herpes viruses recognition quality of. Of course, it should be noticed, that we have the structural parameter α, so we need some instrument for automatic selection of its appropriate value. But this aspect is outside the scope of this investigation.

4 Conclusions and Discussion

This paper proposes a proteins comparison measure based on the optimal global alignment. The proposed measure is very similar to the traditional Needleman-Wunsch similarity measure, but, in contrast to it, possesses metric properties. Also, we have proposed some ways for speeding-up its computation. The obtained results have shown that the use of proposed metric instead of the traditional NW-similarity measure allows for essentially increasing the quality of Herpes viruses recognition.

Acknowledgements. This work is supported by the Russian Foundation for Basic Research, Grant 15-07-08967.

The results of the research project are published with the financial support of Tula State University within the framework of the scientific project - 2017-18PUBL.

References

1. Huleihel, M., Shufan, E., Zeiri, L., Salman, A.: Detection of vero cells infected with Herpes simplex types 1 and 2 and Varicella Zoster viruses using Raman spectroscopy and advanced statistical methods. PLoS ONE **11**(4), e0153599 (2016). https://doi.org/10.1371/journal.pone.0153599
2. Mc Geoch, D.J., Rixon, F.J., Davison, A.J.: Topics in herpesvirus genomics and evolution. Virus Res. **117**, 90–104 (2006). https://doi.org/10.1016/j.virusres.2006.01.002
3. Needleman, S.B., Wunsch, C.D.: A general method applicable to the search for similarities in the amino acid sequence of two proteins. J. Mol. Biol. **48**(3), 443–453 (1970). https://doi.org/10.1016/0022-2836(70)90057-4
4. Smith, T.F., Waterman, M.S.: Identification of common molecular subsequences. J. Mol. Biol. **147**(1), 195–197 (1981). https://doi.org/10.1016/0022-2836(81)90087-5
5. Zhang, Z., Schwartz, S., Wagnerm, L., Miller, W.: A greedy algorithm for aligning DNA sequences. J. Comput. Biol. **7**(1–2), 203–214 (2000). https://doi.org/10.1089/10665270050081478
6. Durbin, R., Eddy, S.R., Krogh, A., Mitchison, G.: Biological Sequence Analysis: Probabilistic Models of Proteins and Nucleic Acids, p. 356. Cambridge University Press, Cambridge (1998)
7. Vapnik, V.N.: Statistical Learning Theory, p. 768. Wiley, Hoboken (1998)
8. Schölkopf, B., Tsuda, K., Vert, J.-P.: Kernel Methods in Computational Biology, p. 410. MIT Press, Cambridge (2004)
9. Aizerman, M.A., et al.: Potential Functions Method in Machine Learning Theory, p. 384. Nauka, Moscow (1970). (in Russian)

10. Sulimova, V.V.: Kernel functions for analysis of signals and symbolic sequences of different length, p. 122. Ph.D. thesis, Tula (2009). (in Russian)
11. Miklós, I., Novak, A., Satija, R., Lyngso, R., Hein, J.: Stochastic models of sequence evolution including insertion-deletion events. Stat. Methods Med. Res. **18**(5), 453–485 (2009). https://doi.org/10.1177/0962280208099500
12. Seeger, M.: Covariance kernels from Bayesian generative models. In: Dietterich, T.G., Becker, S., Ghahramani, Z. (eds.) Advances in Neural Information Processing Systems, vol. 14, pp. 905–912. MIT Press (2002)
13. Abramov, V.I., Seredin, O.S., Mottl, V.V.: Pattern recognition training by support object method in Euclidean metric spaces with affine operations. In: Proceedings of Tula State University. Natural Sciences Series, vol. 2, no. 1, pp. 119–136. TSU, Tula (2013). (in Russian)
14. Pekalska, E.M.: Dissimilarity representations in pattern recognition. Concepts, Theory and Applications. Ph.D. thesis, p. 344 (2005). ISBN 90-9019021-X
15. Seredin O.S., Mottl V.V.: Support object method for pattern recognition training in arbitrary metric spaces. In: Proceedings of Tula State University. Natural Sciences Series, vol. 4, pp. 178–196. TSU, Tula (2015). (in Russian)
16. Braverman, E.M.: Experiments on training a machine for pattern recognition. Ph.D. thesis. Moscow (1961). (in Russian)
17. Xing, E.P., Ng, A.Y., Jordan, M.I., Russel, S.: Distance metric learning with application to clustering with side-information. In: Becker, S., Thrun, S., Obermayer, K. (eds.) Advances in Neural Information Processing Systems, vol. 15, pp. 521–528. MIT Press (2003)
18. Bellet, A., Harbrad, A., Sebban, M.: A survey on metric learning for feature vectors and structured data. CoRR (2013). http://arxiv.org/abs/1306.6709
19. Wang, J., Sun, K., Sha, F., Marchand-Maillet, S., Kalousis, K.: Two-stage metric learning. In: Proceedings of the 31st International Conference on Machine Learning, Cycle 2, vol. 32, pp. 370–378 (2014)
20. Schultz, M., Joachims, T.: Learning a distance metric from relative comparisons. In: Thrun, S., Saul, L.K., Schölkopf, P.B. (eds.) Advances in Neural Information Processing System, vol. 16, pp. 41–48. MIT Press (2004)
21. Wang, J., Do, H., Woznica, A., Kalousis, A.: Metric learning with multiple Kernels. In: Shawe-Taylor, J., Zemel, R. S., Bartlett, P.L., Pereira, F., Weinberger, K.Q. (eds.) Advances in Neural Information Processing Systems, vol. 24, pp. 1–9. Curran Associates, Inc. (2011)
22. Cao, M., Zhang, H., Park, J., Daniels, N.M., Crovella, M.E., et al.: Going the distance for protein function prediction: a new distance metric for protein interaction networks. PLoS ONE **8**(10), e76339 (2013). https://doi.org/10.1371/journal.pone.0076339
23. Rogen, P., Fain, B.: Automatic classification of protein structure by using Gauss integrals. Proc. Natl. Acad. Sci. USA **100**(1), 119–124 (2002). https://doi.org/10.1073/pnas.2636460100
24. Dayhoff, M., Schwarts, R., Orcutt, B.: A model of evolutionary change in proteins. Atlas of Protein Sequences Struct. **5**(3), 345–352 (1978)
25. Mottl, V.V.: Metric spaces admitting linear operations and inner product. Doklady Math. **67**(1), 140–143 (2003)
26. Sulimova, V., Seredin, O., Mottl, V.: Metrics on the basis of optimal alignment of biomolecular sequences. JMLDA **2**(3), 286–304 (2016). https://doi.org/10.21469/22233792.2.3.03

27. Altschul, S.F., Gish, W., Miller, W., Myers, E.W., Lipman, D.J.: Basic local alignment search tool. J. Mol. Biol. **215**(3), 403–410 (1990). https://doi.org/10.1006/jmbi.1990.9999
28. Lipman, D.J., Pearson, W.R.: Rapid and sensitive protein similarity searches. Science **227**(4693), 1435–1441 (1985). https://doi.org/10.1126/science.2983426
29. Pearson, W.R.: Flexible sequence similarity searching with the FASTA3 program package. Methods Mol. Biol. 185–219 (2000). https://doi.org/10.1385/1-59259-192-2:185
30. Sakoe, H., Chiba, S.: Dynamic programming optimization for spoken word recognition. IEEE Trans. Acoust. Speech Signal Process. **26**(1), 43–49 (1978). https://doi.org/10.1109/tassp.1978.1163055
31. Myers, C., Rabiner, L.R., Rosenberg, A.E.: Performance tradeoffs in dynamic time warping algorithms for isolated word recognition. IEEE Trans. Acoust. Speech Signal Process. **28**(6), 623–635 (1980). https://doi.org/10.1109/tassp.1980.1163491
32. Silva, D.F., Batista, G.E.A.P.A.: Speeding up all-pairwise dynamic time warping matrix calculation. In: Proceedings of the 2016 SIAM International Conference on Data Mining, pp. 837–845 (2016). https://doi.org/10.1137/1.9781611974348.94
33. Virus Database at University College London (VIDA). http://www.biochem.ucl.ac.uk/bsm/virus_database/VIDA3/VIDA.html
34. Lanckriet, G., Bie, T.D., Cristianini, N., Jordan, M.I., Noble, W.S.: A statistical framework for genomic data fusion. Bioinformatics **20**(16), 2626–2635 (2004). https://doi.org/10.1093/bioinformatics/bth294

Learning Interpretable Prefix-Based Patterns from Demographic Sequences

Danil Gizdatullin[1][(✉)], Jaume Baixeries[2], Dmitry I. Ignatov[1,4],
Ekaterina Mitrofanova[1], Anna Muratova[1], and Thomas H. Espy[3]

[1] National Research University Higher School of Economics, Moscow, Russia
{dgizdatullin,dignatov,emitrofanova}@hse.ru
[2] Universitat Politècnica de Catalunya, Barcelona, Spain
jbaixer@cs.upc.edu
[3] University of Pittsburgh, Pittsburgh, USA
the7@pitt.edu
[4] St. Petersburg Department of Steklov Mathematical Institute of Russian Academy
of Sciences, Saint Petersburg, Russia

Abstract. There are many different methods for computing relevant patterns in sequential data and interpreting the results. In this paper, we compute emerging patterns (EP) in demographic sequences using sequence-based pattern structures, along with different algorithmic solutions. The purpose of this method is to meet the following domain requirement: the obtained patterns must be (closed) frequent contiguous prefixes of the input sequences. This is required in order for demographers to fully understand and interpret the results.

Keywords: Demographic sequences · Pattern structures · Sequence mining · Emerging patterns · Emerging sequences · Machine learning

1 Introduction and Related Work

Demographic sequences are composed of crucial life course events which occur during transition to adulthood: completing education, getting the first job, family formation etc. The analysis of these sequences is a popular and promising study direction in demography [1,2]. Among different results that can be obtained by the analysis of demographic sequences, in this paper we focus on the computation of events that are relevant for characterising differences across generations. For example, we would like to explore the main differences between consecutive generations, in terms of demographic behaviour: number of children, age of marriage, incidence of divorce, etc.

Demographers and sociologists do not currently have a simple, unified methodology for the computation and interpretation of such event patterns, so different techniques are used: sequence analysis [3–5] and statistical methods [6–10]. Demographers have also started to show great interest in machine-learning and pattern-mining techniques [11] and other sophisticated sequence-analysis techniques [12]. Although many different methods have been developed,

© Springer Nature Switzerland AG 2019
V. V. Strijov et al. (Eds.): IDP 2016, CCIS 794, pp. 74–91, 2019.
https://doi.org/10.1007/978-3-030-35400-8_6

the methodology used in this field is not fully convergent with state-of-the-art sequence-mining techniques.

In the previous paper [13], we used the SPMF (Sequential Pattern Mining Framework) [14] for mining frequent sequences and finding relevant emerging patterns. However, this approach has a drawback: the results it yields are hard to interpret. Our work with demographers made us realize that it would be useful to find contiguous, prefix-based patterns, since they are interested in the full starting parts of people's life trajectories without gaps.

The main goal of this paper is to find emerging patterns that allow us to discern demographic behaviour of different groups of people, with one important restriction, which is necessary to ensure the interpretability of our results: the obtained patterns must be (closed) frequent contiguous prefixes of the input sequences[1].

The paper is organised in a following way. We briefly determine the scope of this paper in Sect. 2. In Sect. 3, we describe our demographic dataset. Section 4 introduces basic definitions and prefix-based contiguous subsequences, in terms of pattern structures, combined with emerging patterns. Experimental results are reported in two subsections of Sect. 5. Finally, in Sect. 6 we provide the main conclusions of this paper.

2   Problem Statement

In general terms, in this paper we compute a set of patterns that can characterise one group of subjects (a generation, a geographically defined group of people, a gender) with respect to other groups. As an example, we would like to be able to answer questions about the relevant differences between men and women or between generations in terms of demographic behaviour or questions about the emerging patterns that distinguish two consecutive generations.

We want to answer all these questions, inter alia, by mining emerging contiguous patterns.

However, it is important to note that we need to compute patterns that can be interpreted by field experts. This is the reason why classification methods like SVM and artificial neural networks must be discarded.

3   Materials

The dataset for the study is obtained from the Research and Educational Group for Fertility, Family Formation and Dissolution at the Higher School of Economics[2]. We use the three wave panel of the Russian part of the Generations

---

[1] Please note this is an extended and updated version of our preliminary studies presented at IDP 2016 conference and ICFCA 2017 [15] especially w.r.t. to a larger dataset used and revisited experiments.

[2] http://www.hse.ru/en/demo/family/.

and Gender Survey (GGS), which took place in 2004, 2007 and 2011[3]. In the preliminary work the dataset contained records of 4,857 respondents (1,545 men and 3,312 women) [15], while now it contains the results of a survey of 6,626 people, including 3,314 men and 3,312 women. This dataset was prepared for testing black-box machine learning approaches in [16] to cope with the imbalance of the original data.

The gender imbalance of the original dataset is caused by the panel nature of the data: survey respondents' attrition is an uncontrollable process. That is why the representative waves combine to form a panel with a structure dissimilar to that of the general sample.

In the dataset, the following information is indicated for each person: date of birth, gender (male, female), generation, type of education (general, professional, higher), locality (city, town, village), religion (yes, no), frequency of religious event attendance (once a week, several times a week, minimum once a month, several times in a year or never). In addition, the dataset provides the dates of significant events in their lives, such as first job experience, completion of education of the highest level, leaving the parental home, first partnership (unregistered union), first marriage and birth of the first child. There are eleven generations: the first is of those born in 1930–1934, the last is of those born in 1980–1984.

4   Sequence Mining and Emerging Patterns

4.1   Pattern Structures in a Demographic Context

A *prefix-based contiguous subsequence* (or simply *prefix*) of a sequence $s = \langle s_1, \ldots, s_k \rangle$ of length $k' \leq k$ is the sequence $s_1 = \langle s'_1, \ldots, s'_{k'} \rangle$, where $s_i = s'_i$ for all $i \leq k'$. The *relative support*, $rsup_T(s)$, of a prefix s in a set of sequences T is the number of sequences in T that start with s divided by $|T|$.

For example, for sequence $\langle \{education\}, \{work\}, \{marriage\} \rangle$, the subsequence $\langle \{education\}, \{marriage\} \rangle$ is not a prefix-based contiguous subsequence. But $\langle \{education\} \rangle$, $\langle \{education\}, \{work\} \rangle$ and $\langle \{education\}, \{work\}, \{marriage\} \rangle$ are in fact prefix-based contiguous subsequences. Pattern structures were introduced in [17] to analyse complex data with object descriptions given in non-object-attribute-value form, for example, chemical graphs, syntactic trees, vectors of numeric intervals and sequences. The usage of pattern structures for sequence mining has already been successfully demonstrated in [18].

Let $(S, (D, \sqcap), \delta)$ be a *pattern structure* related to our demographic problem, where S is a set of sequences, D is a set of possible descriptions of patterns with an associated intersection operator $\sqcap$ and operator $\delta(s)$ returns the description of sequence s from D. For example, if we have two sequences $s_1, s_2 \in S$, then

$$\delta(s_1) = \langle e_1^1, e_2^1, \ldots, e_n^1 \rangle \text{ and } \delta(s_2) = \langle e_1^2, e_2^2, \ldots, e_m^2 \rangle.$$

---

[3] This part of GGS "Parents and Children, Men and Women in Family and in Society" is an all-Russian representative panel sample survey: http://www.ggp-i.org/.

In our case, e_i^j is an event which happened in a person's lifetime.

Given two descriptions $d_1, d_2 \in D$, the intersection operator $\sqcap$ returns their maximal common prefix. To generate the maximal common prefix for a sequences subset, we use the Galois operator denoted by $\diamond$ that results in the intersection of the descriptions of the input sequences. For example, for s_1, s_2 such that $\delta(s_1) = \langle e_1^1, e_2^1, \ldots, e_n^1 \rangle$ and $\delta(s_2) = \langle e_1^2, e_2^2, \ldots, e_m^2 \rangle$,

$$\{s_1, s_2\}^\diamond = \delta(s_1) \sqcap \delta(s_2) = \langle e_1, e_2, \ldots, e_k \rangle$$

where k is maximal such that $e_i = e_i^1 = e_i^2$ for all $i \le k \le \min\{n, m\}$.

The operation $\diamond$ applied to description $d = \langle e_1, e_2, \ldots, e_k \rangle$ in our case is

$$d^\diamond = \{s \in S \mid d \sqsubseteq \delta(s)\},$$

where $d \sqsubseteq \delta(s)$ means that d is a prefix of sequence s ($d \sqsubseteq \delta(s)$ if $d \sqcap \delta(s) = d$). In other words, the operator $\diamond$ applied to a description d returns the subset of objects in S that have d as a prefix.

A pair (A, d) is a *pattern concept* of a pattern structure $(S, (D, \sqcap), \delta)$ with

$$A^\diamond = d \text{ and } d^\diamond = A, \text{ where } A \subseteq S, \text{ and } d \in D.$$

A is called the *extent* of the pattern concept (A, d) and d is its *intent*. For every pattern concept (A, d) of a pattern structure $(S, (D, \sqcap), \delta)$, it follows that $A^{\diamond\diamond} = A$ and $d^{\diamond\diamond} = d$. One may check that $(\cdot)^{\diamond\diamond}$ is a closure operator (idempotent, monotone, extensive) on 2^S w.r.t. $\subseteq$, so A is closed.

Let us discuss the representation of pattern concepts in the related prefix-tree (see Sect. 4.5) for the original pattern structure and how they can be found. As an example, consider the set of sequences S, given by $\delta(s_1) = \langle a, b, c \rangle$, $\delta(s_2) = \langle a, b, c \rangle$ and $\delta(s_3) = \langle a, b, d \rangle$. The corresponding prefix tree is

$$\emptyset$$
$$|$$
$$a(3)$$
$$|$$
$$b(3)$$
$$\overset{\frown}{c(2) \ d(1)}$$

We can extract the following pattern concepts relevant to this example: $(\{s_1, s_2, s_3\}, \langle a, b \rangle)$, $(\{s_1, s_2\}, \langle a, b, c \rangle)$, and $(\{s_3\}, \langle a, b, d \rangle)$. Any path of the associated prefix-tree, from its root to a bottom node, whose support is higher than the support of its descendants corresponds to the concept of an original pattern structure.

4.2 Hypotheses in Pattern Structures

Let us formulate a classification problem in a demographic setting. For each object (individual), there is also a target attribute (e.g. gender, for binary classification) according to which we want to classify that individual. Our pattern structure is then split into two pattern structures, positive $K_\oplus = (S_\oplus, (D, \sqcap), \delta)$ and negative $K_\ominus = (S_\ominus, (D, \sqcap), \delta)$, according to the target attribute which determines the class where it belongs to. Also, we have a set of undetermined sequences S_τ with unknown target attribute value.

Now, when the associated Galois operator is denoted as $A^\oplus$ for the positive pattern context and correspondingly for the negative one.

Let us define *positive* and *negative hypotheses*. A pattern intent of the pattern structure $K_\oplus$ ($K_\ominus$) $H \sqsubseteq D$ is a positive (negative) hypothesis if H is not a subset of the pattern intent of any negative (positive) examples $s \in S_\ominus$ ($s \in S_\oplus$):

$$\forall s \in S_\ominus (s \in S_\oplus) : H \not\sqsubseteq s^\ominus (H \not\sqsubseteq s^\oplus).$$

Eventually, the hypothesis is the pattern intent of a pattern concept, which is found only in the objects of just one class.

4.3 Emerging Patterns in Pattern Structures

Also, we introduce the notion of *emerging prefix-based contiguous subsequences* in terms of pattern structures. *Emerging pattern* is specific for one class, but not specific for its counterpart.

This feature is implemented via the ratio of the pattern supports for different classes. This ratio is called *growth rate*. The growth rate of a pattern $p \in D$ on positive and negative pattern structures of $K_\oplus$ and $K_\ominus$ is defined as

$$GR(p, K_\oplus, K_\ominus) = \frac{rsup_{K_\oplus}(p)}{rsup_{K_\ominus}(p)}$$

Patterns are selected by specifying a minimum growth rate as in [19]. That means, we set the minimum growth ratio, for which we want to select patterns:

$$GR(p, K_\oplus, K_\ominus) \geq \theta$$

Let us consider an example. Assume that we have two sets of sequences. Men's sequences:

$$\langle \{education\}, \{work\}, \{marriage\} \rangle$$
$$\langle \{education\}, \{work\}, \{marriage\} \rangle$$
$$\langle \{education\}, \{marriage\}, \{work\} \rangle$$

Women's sequences:

$$\langle \{education\}, \{marriage\}, \{work\} \rangle$$
$$\langle \{marriage\}, \{education\}, \{work\} \rangle$$
$$\langle \{marriage\}, \{education\}, \{work\} \rangle$$

So we can make a prefix-tree which based on these data.

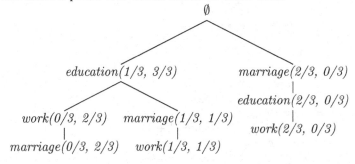

On each node of the tree we store *rsup* for different classes of current prefix-based contiguous subsequences. By using this structure we can compute growth rate.

4.4   Usage of Emerging Patterns for Classification

For each class we compute its score as suggested in [20]. Let s be a new sequence which we want to classify, then its score in positive class is equal to

$$score_\oplus(s) = \frac{\sum_{p \in D^\theta_\oplus, p \sqsubseteq \delta(s)} GR(p, K_\oplus, K_\ominus)}{median(GR(S_\oplus))}$$

where $D^\theta_\oplus$ is a subset of D that consists of all p with $GR(p, K_\oplus, K_\ominus) \geq \theta$. Then we choose all prefixes for the new sequence from set $D^\theta_\oplus$, and we sum all these growth-rates and get the score. Then we normalize the score by the median of the growth-rate for the current (positive) class.

4.5   Prefix-Tree Building: Pseudocode and Complexity

As we need to find contiguous prefix subsequences, it is a good idea to use a prefix-tree [21]. Usually, every node in a prefix-tree is associated with a certain string, but in our tree structure each node is associated with only one symbol (in case there are no simultaneous events).

Prefix-Tree Building. Let us start with the prefix-tree building procedure. As an input we have the set of sequences and their labels. At first we create the root node, which should be empty. Then we iterate through all the sequences and try to go the full path from the root to the end of a sequence. If we encounter a new path, we create a new node (and all the remaining events in the current sequence should be added as new subsequent nodes). Along the way, we increment all the counters associated with traversed nodes, one for each class.

Time Complexity. Let n be the size of the training set and m be the number of different events in it. In line 4, we go through all the data; it takes n times.

Algorithm 1. Sequence tree building

1: **procedure** SEQUENCESTREE(S, L)
2:      $T \leftarrow \{\emptyset\}$ // Initial prefix tree
3:      $cn \leftarrow \emptyset$
4:      **for** $s \in S$ **do**
5:          $l \leftarrow label(s)$
6:          **for** $e \in s$ **do**
7:              $c \leftarrow$ FIND(e, cn.children)
8:              **if** $c \neq \emptyset$ **then**
9:                  c.counter$[l] \leftarrow c$.counter$[l] + 1$
10:                  $cn \leftarrow c$
11:              **else**
12:                  cn.children.append($newC$)
13:                  $newC$.element $\leftarrow e$
14:                  $newC$.counter$[l] \leftarrow 1$
15:                  $cn \leftarrow newC$

1: **function** FIND(e, N)
2:      **for** $n \in N$ **do**
3:          **if** n.element $= e$ **then return** n
     return None

Then, in line 6, we go through all the elements in a sequence. The maximum length of a sequence is m. This pass takes $O(m)$ steps. Then in FIND procedure we iterate through all the children of a node. The maximum number of children nodes is $m-1$; this step takes $m-1$. Thus, the total time complexity is $O(n \cdot m^2)$. In our case, m is a small value (7–10 events) and can be considered a constant. The time complexity is $O(n)$.

Space Complexity. The space complexity is equal to the number of nodes in a tree. The worst case is when all n sequences are different, i.e. all n sequences do not have the same prefix. In this case, space complexity will be $O(n \cdot m)$.

Classification by Patterns. After performing the SEQUENCESTREE procedure on input data, we have the prefix-tree with absolute support value for each node and label. Then we can classify a new portion of sequences from the same domain. At first, we perform preprocessing via the PRECOMPUTEGROWTHRATE procedure. In this procedure, we compute relative support and growth rates for each node. After that, we use the CLASSIFYSEQUENCE function to predict the label of a new sequence.

Time Complexity. Let k be the length of a sequence for classification. In PRECOMPUTEGROWTHRATE, we need to iterate through the tree's nodes two times for each class label. We consider the situation with only two different classes: $O(n \cdot m \cdot 2) = O(n \cdot m)$.

In CLASSIFYSEQUENCE, we iterate through the elements of the sequence and node children of nodes for each label: $O(k \cdot m \cdot 2) = O(k \cdot m)$.

Algorithm 2. Classify Sequence

1: **function** CLASSIFYSEQUENCE($T, s, l, Classes, minSup, minGR$)
2:     $sfc \leftarrow [0, 0]$
3:     $cn \leftarrow T.\text{root}$
4:     **for** $e \in s$ **do**
5:         **for** $c \in cn.\text{children}$ **do**
6:             **if** $c.\text{element} = e$ **then**
7:                 **for** $l \in Classes$ **do**
8:                     **if** $(c.support[l] > minSup)$ **and** $(c.GR[label] > minGR)$ **then**
9:                         $sfc[l] \leftarrow sfc[l] + n.GR[l]$
10:                        $cn \leftarrow c$
     return $argMax(sfc)$

1: **procedure** PRECOMPUTEGROWTHRATE($T, Classes, soc$)
2:     $soc \leftarrow \text{size}(Classes)$ // soc is the number of classes
3:     **for** $n \in T$ **do** // iterate over the tree nodes
4:         **for** $l \in Classes$ **do** // iterate over the labels of classes
5:             $n.\text{support}[l] \leftarrow n.\text{counter}[l]/\text{soc}[l]$
6:     **for** $n \in T$ **do**
7:         **for** $l \in C$ **do** // GR is a growth-rate attribute
8:             $n.GR[l] \leftarrow n.\text{support}[l]/n.\text{support}[counterpartL]$

5   Experiments and Results

To perform experiments with pattern-based classification, we use Python and the Contiguous Sequences Analysis library implemented by the first author[4].

5.1   Classification by Gender

After discussing with demographers, we have set the minimal relative support at 0.03. We have received the following prefix-based contiguous patterns that meet a minimum of 3% of all respondents in Tables 1 and 2.

After thoughtful inspection, we can conclude that the beginning of human life trajectories do not depend strongly on gender; moreover, the beginnings of the most popular paths are the same for both sexes. We have split all our data into two groups: a training set and a test set with the percentage of 66.5%–33.5%.

We have performed simple gridsearch over two sets of parameters, $minsup \in \{0.001, 0.004, 0.01, 0.025, 0.04, 0.05, 0.1\}$ and for $min\_growth\_rate \in \{1.5, 1.25, 2, 2.25, 3, 3.5, 5, 7\}$. The results are summarised in Table 3[5].

The graphs below show the results and skyline in TPR-FPR (true positive rate, false positive rate), TPR-NCPR (true positive rate, non-classified positive

---

[4] https://github.com/DanilGizdatullin/ContiguousSequencesAnalysis.

[5] Note that $NCR \approx 0.996$ means that 96.6% of objects remain without classification. Together with $TPR = 0$ and $FPR = 0$ this implies that we deal only with objects of the negative class. In this case, $TNR = 7$, $FNR = 2$ means 7 out of 9 objects are correctly classified with $Accuracy \approx 0.78$.

Table 1. Women's patterns

Pattern	Support
$\langle\{work\}\rangle$	0.287
$\langle\{work\}, \{education\}\rangle$	0.120
$\langle\{separation\}\rangle$	0.283
$\langle\{education\}\rangle$	0.239
$\langle\{education\}, \{work\}\rangle$	0.168
$\langle\{separation\}, \{education\}\rangle$	0.110
$\langle\{separation\}, \{education\}, \{work\}\rangle$	0.097

Table 2. Men's patterns

Pattern	Support
$\langle\{work\}\rangle$	0.329
$\langle\{work\}, \{education\}\rangle$	0.155
$\langle\{separation\}\rangle$	0.266
$\langle\{education\}\rangle$	0.276
$\langle\{education\}, \{work\}\rangle$	0.103
$\langle\{separation\}, \{education\}\rangle$	0.199
$\langle\{separation\}, \{education\}, \{work\}\rangle$	0.099

rate), NCPR-FPR (non-classified positive rate, false positive rate) on the axes (Figs. 1 and 2).

We have chosen minimum support values as 0.003 for women and 0.002 for men, respectively, while the original minimum support has been set as 0.001 for all the rules. We have performed several classifications with different minimum values of the growth rate from $\{1.5, 2, 2.25, 3, 5, 7\}$. In a demographic setting, it is important to identify interesting discriminative patterns, but not only solve the problem of classification by gender. Thus, many objects from the test set have not been assigned to any class. For example, in the experiment with the best combination of TPR and FPR metrics, we cover over 6% of people in the test sample. Sample. We can conclude from the obtained results that many interesting discriminative patterns for some class relative to another have a small cover. Moreover, we can conclude that the average behaviour of men and women has no stark differences in general, but there are local groups of both classes that behave sufficiently differently.

The best quality of classification has been reached with the minimum value of the growth rate 7. It corresponds to the following emerging patterns in Tables 4 and 5, respectively.

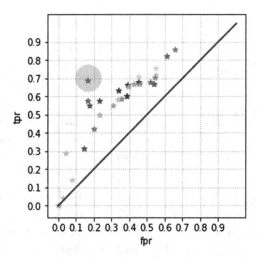

Fig. 1. TPR-FPR plot, along with one Pareto-optimal result with high TPR from the skyline in the oval.

The results show that women very often start their transition to adulthood with marital events: in two the most frequent sequences, marriage is the first event; in two sequences – partnership is the third or the fourth event[6].

All the frequent sequences of men's biographies contain a marital event only at the fourth place, and all their first three events are socioeconomic. It means that men are more oriented towards career building and financial independence while some women tend to create a family first. For such women, the prioriti-

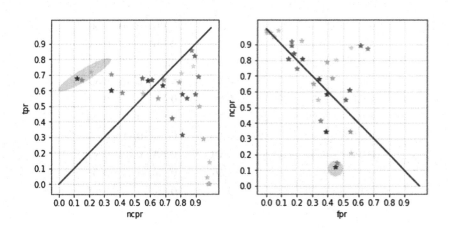

Fig. 2. TPR-NCPR (left) and NCPR-FPR (right) plots, along with some Pareto-optimal results from their skylines in the ovals.

---

[6] Dissolution is a short form for "dissolution of partnership" event.

Table 3. Gridsearch over of minimal support and minimal growth rate values

Minsup	MinGrowthRate	Accuracy	TPR	FPR	NCR
0.001	1.5	0.63	**0.67**	0.42	0.26
0.001	1.25	0.60	0.65	0.44	**0.01**
0.001	2	0.64	0.61	0.33	0.53
0.001	2.25	0.63	0.56	0.30	0.67
0.001	3	0.66	0.63	0.32	0.81
0.001	3.5	0.69	0.57	0.21	0.87
0.001	5	0.72	0.55	0.17	0.91
0.001	7	**0.78**	**0.67**	**0.14**	0.94
0.004	1.5	0.61	0.66	0.44	0.33
0.004	1.25	0.59	**0.67**	0.49	**0.05**
0.004	2	0.63	0.61	0.34	0.63
0.004	2.25	0.63	0.58	0.32	0.79
0.004	3	0.68	0.64	0.29	0.93
0.004	3.5	0.70	0.34	0.04	0.96
0.004	5	**0.75**	0.00	**0.00**	0.98
0.004	7	0.70	0.00	**0.00**	0.99
0.01	1.5	0.62	0.68	0.45	0.44
0.01	1.25	0.58	0.67	0.52	**0.10**
0.01	2	0.65	0.61	0.30	0.74
0.01	2.25	0.64	0.58	0.29	0.85
0.01	3	0.66	0.71	0.39	0.96
0.01	3.5	0.62	0.00	0.00	0.99
0.01	5	**0.78**	0.00	**0.00**	1.00
0.025	1.5	0.62	**0.69**	0.46	0.64
0.025	1.25	0.57	0.68	0.53	**0.19**
0.025	2	**0.68**	0.77	**0.45**	0.83
0.04	1.5	0.62	0.60	**0.36**	0.75
0.04	1.25	0.57	0.68	0.53	**0.24**
0.04	2	**0.66**	0.72	0.41	0.85
0.05	1.5	**0.61**	**0.55**	**0.34**	0.82
0.05	1.25	0.56	0.75	0.64	**0.35**
0.1	1.25	0.55	0.85	0.80	0.45

zation of family formation leads to delaying of their professional careers (and, maybe, to a complete absence of it). The different place of education in the order of events in men's and women's biographies may also mean the different levels

Table 4. Top-5 women's patterns in the test set for $minsup = 0.003$

Pattern	Growth rate	Support
$\langle\{marriage\}, \{separation\}, \{children\}, \{education\},$ $\{work\}, \{divorce\}\rangle$	∞	0.0036
$\langle\{work\}, \{separation\}, \{partner\}, \{marriage\},$ $\{children\}, \{education\}, \{dissolution\}\rangle$	∞	0.0032
$\langle\{separation\}, \{education\}, \{work\}, \{marriage\},$ $\{children\}, \{divorce\}, \{partner\}, \{dissolution\}\rangle$	∞	0.0036
$\langle\{marriage\}, \{education\}, \{separation\}\rangle$	∞	0.0041
$\langle\{separation\}, \{education\}, \{work\}, \{partner\},$ $\{dissolution\}, \{children\}\rangle$	∞	0.0032

Table 5. Top-3 men's patterns in the test set for $minsup = 0.002$

Pattern	Growth rate	rsup
$\langle\{education\}, \{separation\}, \{work\}, \{marriage\}, \{divorce\}\rangle$	∞	0.003
$\langle\{education\}, \{separation\}, \{work\}, \{partner\}, \{dissolution\}\rangle$	∞	0.0028
$\langle\{separation\}, \{education\}, \{work\}, \{partner\}, \{marriage\},$ $\{dissolution\}, \{divorce\}\rangle$	∞	0.0023

of their education: the higher the age of completing education, the more likely it is of a higher level.

Separation from parental home is the most united event of men's and women's sequences, but it has different meaning depending on in combination with which events it occurs. For women, the driven force of separation from parents is often a partnership or a marriage, while for men, separation is primarily connected either with education or work.

Childbirth is the most dividing event by gender: almost all the frequent sequences of women contain childbirth in the middle or in the end of the transition to adulthood while men demonstrate a complete absence of this event.

5.2   Classification by Generation

In this experiment we search for emerging patterns for different generations of the same sex. The first class 0 features people who were born between 1924 and 1959. The second class 1 contains people who were born between 1960 and 1984.

First, let us find emerging patterns for women from different generations. We have 1,926 women from class 0 and 1,386 women from class 1. We need to tune two parameters: the first is the minimal support and the second is minimal growth rate.

Let us tune the minimal support parameter (Table 6).

As we can see, the minimal support can sufficiently change only the non-classification rate and slightly affect accuracy, the TPR and the FPR.

Table 6. Tuning of minimal support for women

minsup	Accuracy	TPR	FPR	NCR
0.001	0.62	0.49	0.30	**0.20**
0.004	0.63	**0.51**	0.29	0.25
0.01	0.63	0.50	0.28	0.28
0.025	0.65	0.47	0.22	0.38
0.04	0.66	0.43	0.19	0.47
0.05	0.66	0.39	0.17	0.49
0.1	**0.70**	0.00	**0.00**	0.69

We have chosen 0.004 as the minimal support and tuned minimal growth rate (Table 7).

Table 7. Tuning of minimal growth rate for women

minGrowthRate	Accuracy	TPR	FPR	NCR
1.5	0.63	0.51	0.29	**0.25**
2	0.68	0.55	**0.24**	0.51
2.25	**0.71**	0.61	**0.24**	0.63
3	0.69	0.68	0.30	0.78
5	0.70	0.81	0.40	0.91
7	0.69	**0.85**	0.44	0.96

We have decided to choose a minGrowthRate of 2.25, since it covers 0.37% of the test data and provides good results in terms of accuracy, TPR, and FPR.

Since we have had many emerging patterns in the data, we consider only patterns with the greatest growth rate and support.

The difference between women of different generations is very pronounced (Tables 8–9). Almost all the frequent sequences of older generations begin with work or education. It makes older women look much similar to men (Table 5) rather than women in general (Table 4). We need to remember that the most frequent full sequences represent only a small share of the sample – not all the respondents, so the full sequences could be not the most typical.

The sequences of younger women show that partnership now is the most popular and necessary starting event for a woman. We do not see such long sequences as in the older generations' biographies because these women are still going through the process of the transition to adulthood, so they have not had enough time yet to obtain such events as childbearing or higher education.

Let us find emerging patterns for men from different generations. Again we should tune the minimal support (Table 10):

Table 8. Patterns for women of older generations

Pattern	Growth rate	Support
$\langle\{work\}, \{separation\}, \{education\}, \{children\}\rangle$	8.1	0.01
$\langle\{work\}, \{education\}, \{separation\}, \{children\}\rangle$	7.3	0.01
$\langle\{work\}, \{education\}, \{marriage\}, \{children\}, \{separation\}\rangle$	5.9	0.02
$\langle\{work\}, \{marriage\}, \{separation\}, \{education\}, \{children\}\rangle$	5.5	0.01
$\langle\{education\}, \{work\}, \{separation\}, \{marriage\}\rangle$	4.6	0.04
$\langle\{education\}, \{work\}, \{separation\}, \{marriage\}, \{children\}\rangle$	4.4	0.04
$\langle\{work\}, \{marriage\}, \{separation\}, \{education\}\rangle$	4.1	0.01
$\langle\{work\}, \{education\}, \{marriage\}, \{children\}\rangle$	3.8	0.025
$\langle\{education\}, \{separation\}, \{work\}, \{marriage\}, \{children\}\rangle$	3.4	0.01
$\langle\{work\}, \{separation\}, \{education\}, \{marriage\}\rangle$	3.3	0.02

Table 9. Patterns for women of younger generations

Pattern	Growth rate	Support
$\langle\{partner\}, \{separation\}, \{work\}\rangle$	∞	0.01
$\langle\{marriage\}, \{separation\}, \{education\}, \{children\}\rangle$	17.7	0.01
$\langle\{marriage\}, \{separation\}, \{education\}, \{children\}, \{work\}\rangle$	15.0	0.01
$\langle\{partner\}, \{marriage\}, \{children\}\rangle$	13.6	0.01
$\langle\{partner\}, \{education\}, \{work\}\rangle$	12.3	0.01
$\langle\{partner\}, \{education\}\rangle$	9.55	0.015
$\langle\{partner\}, \{work\}\rangle$	6.1	0.01
$\langle\{partner\}\rangle$	5.2	0.09
$\langle\{partner\}, \{marriage\}\rangle$	5.0	0.02
$\langle\{work\}, \{partner\}, \{marriage\}, \{separation\}\rangle$	4.8	0.01

Table 10. Tuning of minimal support for men

minsup	Accuracy	TPR	FPR	NCR
0.001	**0.70**	**0.59**	0.18	**0.14**
0.004	0.69	0.58	0.20	0.17
0.01	0.68	0.53	0.17	0.23
0.025	0.67	0.48	0.14	0.32
0.04	0.66	0.34	**0.07**	0.53
0.05	0.66	0.34	**0.07**	0.53

Again, minimal support can sufficiently change only the non-classification rate.

We have chosen 0.001 as the minimal support and tuned minimal growth rate (Table 11).

Table 11. Tuning of the minimal growth rate for men

minGrowthRate	Accuracy	TPR	FPR	NCR
1.5	0.70	0.59	0.18	**0.14**
2	0.76	0.73	0.21	0.35
2.25	0.77	0.74	0.20	0.45
3	0.83	0.82	0.16	0.61
5	0.90	0.90	**0.10**	0.76
7	**0.92**	**0.94**	0.11	0.80

The patterns with the biggest growth rate and support are reported in Tables 12 and 13.

As in the previous experiment with the women subsample, the main difference lies in the tendency to obtain education; thus, men of younger generation demonstrates this tendency.

Table 12. Patterns for men of older generations

Pattern	Growth rate	Support
$\langle\{work\}, \{marriage\}, \{children\}, \{education\}, \{separation\}\rangle$	8.3	0.02
$\langle\{work\}, \{marriage\}, \{children\}, \{education\}\rangle$	5.4	0.02
$\langle\{work\}, \{separation\}, \{marriage\}, \{children\}, \{education\}\rangle$	4.4	0.03
$\langle\{work\}, \{marriage\}, \{children\}\rangle$	3.8	0.04
$\langle\{work\}, \{separation\}, \{marriage\}\rangle$	3.7	0.05
$\langle\{separation\}, \{work\}, \{education\}, \{marriage\}, \{children\}\rangle$	3.3	0.03
$\langle\{separation\}, \{marriage\}, \{education\}\rangle$	3.3	0.02
$\langle\{separation\}, \{work\}, \{education\}, \{marriage\}\rangle$	3.2	0.04
$\langle\{work\}, \{education\}, \{marriage\}, \{children\}, \{separation\}\rangle$	2.7	0.04
$\langle\{work\}, \{marriage\}\rangle$	2.48	0.09

Table 13. Patterns for men of younger generations

Pattern	Growth rate	Support
$\langle\{education\}, \{work\}, \{partner\},$ $\{separation\}, \{marriage\}, \{children\}\rangle$	∞	0.02
$\langle\{education\}, \{work\}, \{partner\}, \{separation\}, \{marriage\}\rangle$	23.0	0.02
$\langle\{education\}, \{work\}, \{partner\}, \{separation\}\rangle$	4.9	0.04
$\langle\{education\}, \{marriage\}\rangle$	3.1	0.04
$\langle\{work\}, \{education\}, \{partner\}\rangle$	3.0	0.04
$\langle\{partner\}\rangle$	2.5	0.03
$\langle\{education\}, \{work\}, \{separation\}, \{partner\}\rangle$	2.3	0.03

6 Conclusion

The main result of our work is the application of various pattern-mining approaches, including pattern structures, to the analysis of demographic sequences. The following conclusions can be drawn from the first results of this work:

1. In this paper, the application of sequence-based patterns for problems of demographic trajectories has been studied.
2. A new method based on pattern structures for the analysis of special pattern type required by demographers (prefix-based and contiguous) has been proposed and implemented.
3. Behavior patterns for different classes of respondents were obtained and interpreted for the most recent and clean demographic material available for Russia.
4. Classifications based on pattern structures and emerging patterns have been designed and tested.

According to the demographers involved in the project, the work is very important for the further development of the pattern-mining application for demographic analysis of sequence data. Thus, among the next planned steps are the following:

- using similarity [22] and kernel-based approaches [23] for demographic-sequence mining (this is partially fulfilled at the moment in [16]);
- (sub)sequence clustering, in particular, based on pattern structures;
- pattern-mining and rule-based approaches for next-event prediction [13] competitive with black-box approaches like recurrent neural networks (this is partially fulfilled at the moment in [16]);
- comprehensive trajectory visualisation within cohorts [24];
- analysing sequences of statuses like $\langle\{studying, single\}, \{working, single\}\rangle$;
- analysis of matrimonial and reproductive biographies, migration studies, etc.

Acknowledgments. We would like to thank our colleagues Sergei Kuznetsov, Alexey Buzmakov, and Mehdi Kaytoue for their advice on pattern structures and sequence-mining, as well as Guozhu Dong for the interest in our work with emerging patterns and our colleagues from Institute of Demography at the National Research University Higher School of Economics.

This article was prepared within the framework of the Academic Fund Program at the National Research University Higher School of Economics (HSE) in 2016-2017 (grant no. 16-05-0011 "Development and testing of demographic sequence analysis and mining techniques') and by the Russian Academic Excellence Project "5-100".

The work of Dmitry Ignatov on final updates requested by Springer's representatives (Sects. 3, 6, and 5) was supported by the Russian Science Foundation under grant no. 17-11-01276 and performed at St. Petersburg Department of Steklov Mathematical Institute of Russian Academy of Sciences, Russia.

References

1. Aisenbrey, S., Fasang, A.E.: New life for old ideas: the 'second wave' of sequence analysis bringing the 'course' back into the life course. Sociol. Methods Res. **38**(3), 420–462 (2010)
2. Billari, F.C.: Sequence analysis in demographic research. Can. Stud. Popul. **28**(2), 439–458 (2001)
3. Aassve, A., Billari, F.C., Piccarreta, R.: Strings of adulthood: a sequence analysis of young british women's work-family trajectories. Eur. J. Popul. **23**(3/4), 369–388 (2007)
4. Braboy Jackson, P., Berkowitz, A.: The structure of the life course: gender and racioethnic variation in the occurrence and sequencing of role transitions. Adv. Life Course Res. **9**, 55–90 (2005)
5. Worts, D., Sacker, A., McMunn, A., McDonough, P.: Individualization, opportunity and jeopardy in American women's work and family lives: a multi-state sequence analysis. Adv. Life Course Res. **18**(4), 296–318 (2013)
6. Abbott, A., Tsay, A.: Sequence analysis and optimal matching methods in sociology: review and prospect. Sociol. Methods Res. **29**, 3–33 (2000)
7. Billari, F., Piccarreta, R.: Analyzing demographic life courses through sequence analysis. Math. Popul. Stud. **12**(2), 81–106 (2005)
8. Billari, F.C., Fürnkranz, J., Prskawetz, A.: Timing, sequencing, and quantum of life course events: a machine learning approach. Eur. J. Popul. **22**(1), 37–65 (2006)
9. Gauthier, J.A., Widmer, E.D., Bucher, P., Notredame, C.: How much does it cost? Optimization of costs in sequence analysis of social science data. Sociol. Methods Res. **38**(1), 197–231 (2009)
10. Ritschard, G., Oris, M.: Life course data in demography and social sciences: statistical and data-mining approaches. Adv. Life Course Res. **10**, 283–314 (2005)
11. Blockeel, H., Fürnkranz, J., Prskawetz, A., Billari, F.C.: Detecting temporal change in event sequences: an application to demographic data. In: De Raedt, L., Siebes, A. (eds.) PKDD 2001. LNCS (LNAI), vol. 2168, pp. 29–41. Springer, Heidelberg (2001). https://doi.org/10.1007/3-540-44794-6_3
12. Gabadinho, A., Ritschard, G., Müller, N.S., Studer, M.: Analyzing and visualizing state sequences in R with TraMineR. J. Stat. Softw. **40**(4), 1–37 (2011)

13. Ignatov, D.I., Mitrofanova, E., Muratova, A., Gizdatullin, D.: Pattern mining and machine learning for demographic sequences. In: Klinov, P., Mouromtsev, D. (eds.) KESW 2015. CCIS, vol. 518, pp. 225–239. Springer, Cham (2015). https://doi.org/10.1007/978-3-319-24543-0_17

14. Fournier-Viger, P., et al.: The SPMF open-source data mining library version 2. In: Berendt, B., et al. (eds.) ECML PKDD 2016. LNCS (LNAI). Part III, vol. 9853, pp. 36–40. Springer, Cham (2016). https://doi.org/10.1007/978-3-319-46131-1_8

15. Gizdatullin, D., Ignatov, D., Mitrofanova, E., Muratova, A.: Classification of demographic sequences based on pattern structures and emerging patterns. In: Supplementary Proceedings of 14th International Conference on Formal Concept Analysis, ICFCA 2017, Rennes, France, 13–16 June 2017, pp. 49–66 (2017)

16. Muratova, A., Sushko, P., Espy, T.H.: Black-box classification techniques for demographic sequences: from customised SVM to RNN. In: Fourth International Workshop on Experimental Economics and Machine Learning, EEML 2017, Dresden, Germany, 17–18 September 2017, vol. 1968, pp. 31–40. CEUR-WS Proceedings (2017)

17. Ganter, B., Kuznetsov, S.O.: Pattern structures and their projections. In: Delugach, H.S., Stumme, G. (eds.) ICCS-ConceptStruct 2001. LNCS (LNAI), vol. 2120, pp. 129–142. Springer, Heidelberg (2001). https://doi.org/10.1007/3-540-44583-8_10

18. Buzmakov, A., Egho, E., Jay, N., Kuznetsov, S.O., Napoli, A., Raïssi, C.: On projections of sequential pattern structures (with an application on care trajectories). In: 10th International Conference on Concept Lattices and Their Applications, pp. 199–208 (2013)

19. Dong, G., Zhang, X., Wong, L., Li, J.: CAEP: classification by aggregating emerging patterns. In: Arikawa, S., Furukawa, K. (eds.) DS 1999. LNCS (LNAI), vol. 1721, pp. 30–42. Springer, Heidelberg (1999). https://doi.org/10.1007/3-540-46846-3_4

20. Dong, G., Li, J.: Efficient mining of emerging patterns: discovering trends and differences. In: Proceedings of the Fifth ACM SIGKDD International Conference on Knowledge Discovery and Data Mining, KDD 1999, pp. 43–52. ACM (1999)

21. de la Briandais, R.: File searching using variable length keys. In: Proceedings of Western Joint Computer Conference, pp. 295–298 (1959). Cited by Brass. Rene

22. Egho, E., Raïssi, C., Calders, T., Jay, N., Napoli, A.: On measuring similarity for sequences of itemsets. Data Min. Knowl. Discov. **29**(3), 732–764 (2015)

23. Elzinga, C.H., Wang, H.: Versatile string kernels. Theoret. Comput. Sci. **495**, 50–65 (2013)

24. Jensen, A.B., et al.: Temporal disease trajectories condensed from population-wide registry data covering 6.2 million patients. Nat. Commun. **5** (2014). Article no. 4022

Population Health Assessment Based on Entropy Modeling of Multidimensional Stochastic Systems

Alexander N. Tyrsin[1,2] and Garnik G. Gevorgyan[1(✉)]

[1] Ural Federal University, Yekaterinburg, Russia
at2001@yandex.ru, garnik.ggg@gmail.com
[2] Scientific and Engineering Center Reliability and Resource of Large Systems
and Machine of UB RAS, Yekaterinburg, Russia

Abstract. The article attempts to demonstrate the possible application opportunities of multidimensional stochastic systems entropy modeling in medicine on a specific example. An assessment of rural male population health state has been done by the criteria of "healthy", "practically healthy", and "diseased". The entropy levels are determined for the parameters, characterizing the main risk factors for chronic noncommunicable diseases, depending on the health group. For each group a single parameter with the largest contribution to the entropy of the population is highlighted, and the dependence degree of this parameter from the others is determined. It was concluded that with the deterioration of health state the population entropy increases. This growth is caused by the increase of randomness entropy. It is shown that with the help of the entropy model it was possible to quite confidently study population health simultaneously by several risk factors. This allows us to count on the successful application of the proposed approach in medicine, as well as in other spheres.

Keywords: Entropy · Model · Multidimensional random variable · Variance · Coefficient of determination · Risk factor · Population · Health state

1 Introduction

One of the least-studied and significant problems in medicine is the comprehensive assessment of population health by multiple risk factors and their relationship. The reason is that the contribution of each risk factor to the overall health state assessment is not that obvious. Various methods are known for studying the health state. The expert assessments, often used in such cases, cannot be considered fully objective [6,13]. Human biological parameters are characterized by high variability, instability, and mostly by non-gaussian distribution, nonlinearity of correlation dependencies, which significantly complicates the regression construction between health state and risk factors. In addition, there are no

© Springer Nature Switzerland AG 2019
V. V. Strijov et al. (Eds.): IDP 2016, CCIS 794, pp. 92–105, 2019.
https://doi.org/10.1007/978-3-030-35400-8_7

valid scalar quantitative estimates of human health state, and the application of regression models with qualitative dependent variables does not allow an explicit consideration of a relationship between risk factors.

The use of a variance analysis requires some risk factor levels management, which is practically unrealistic in the population health study.

The survival analysis, widely utilized in medical studies, is used to evaluate life expectancy while studying the treatment methods effectiveness and does not allow to study population health simultaneously by several risk factors.

Therefore, it seems to be an urgent problem to create methods that allow us to examine human health state in a comprehensive manner, taking into account simultaneously many different interconnected risk factors.

One of the promising alternative approaches to the comprehensive assessment of population health could be the entropy modeling [15]. Entropy is a fundamental property of all systems with an ambiguous or probabilistic behavior [7]. The concept of entropy is flexible and it can be clearly interpreted in terms of that specific science section it is applied to. It is being widely used in modern science to describe the structural organization and disorganization, the destruction connections degree between the system elements. Therefore, it seems that the entropy could act as a universal parameter and it is ideal for solving the behavior problems of complex stochastic systems [12]. Each and every object and phenomena of nature and wildlife, contains traits of order and disorder, certainty and uncertainty, organization and disorganization.

The paper aims to investigate the entropy approach possibilities for population health assessment. To this end, on the basis of experimental data let us build the rural male health entropy model simultaneously for several risk factors and examine the influence of these factors on population health.

2 The Differential Entropy of Multidimensional Random Variables

The differential entropy of a multidimensional continuous random variable $Y = (Y_1, \ldots, Y_m)$ was introduced by Shannon in 1948 [10]. It is defined as:

$$H(Y) = -\int_{-\infty}^{+\infty} \ldots \int_{-\infty}^{+\infty} \ln p_y(x_1, \ldots x_m) dx_1 \ldots dx_m,$$

where $p_y(x_1, \ldots x_m)$ is the joint probability density function (PDF) of the random variables $Y_1, \ldots, Y_m$.

In [15] it is proved that if all the components of Y_i have variances $\sigma_{Y_i}^2$, then the differential entropy (the entropy from now on) $H(Y)$ of the random variable Y is equal to:

$$H(Y) = \sum_{i=1}^{m} \ln \sigma_{Y_i} + \sum_{i=1}^{m} \kappa_i + \tfrac{1}{2} \sum_{k=2}^{m} \ln(1 - R_{Y_k/Y_1 \ldots Y_{k-1}}^2), \tag{1}$$

where $\kappa_i = H(Y_i/\sigma_{Y_i})$ is the entropy indicator of the distribution type of Y_i random variable; $R^2_{Y_k/Y_1\ldots Y_{k-1}}$ are the determination coefficients of the regression dependencies; $k = 2, \ldots, m$.

The first two terms $H(\boldsymbol{Y})_V = \sum_{i=1}^m H(Y_i) = \sum_{i=1}^m \ln \sigma_{Y_i} + \sum_{i=1}^m \kappa_i$ are the entropy of a random vector with mutually independent components and are called the randomness entropy, the third one – $H(\boldsymbol{Y})_R = \frac{1}{2}\sum_{k=2}^m \ln(1 - R^2_{Y_k/Y_1\ldots Y_{k-1}})$—self-organization entropy.

Note that if the random vector $\boldsymbol{Y}$ is Gaussian, then we have a particular case, where $H(\boldsymbol{Y})_V = \sum_{i=1}^m H(Y_i) = \sum_{i=1}^m \ln \sigma_{Y_i} + m \ln \sqrt{2\pi e}$, $H(\boldsymbol{Y})_R = \frac{1}{2}\ln |R|$, and $\boldsymbol{R}$ is the correlation matrix of random vector $\boldsymbol{Y}$.

According to (1), the $H(\boldsymbol{Y})$ entropy possesses trialism. There are three reasons for its change: the change of components scattering level, the change of components distribution type, and the change of correlation strength between its components.

The Eq. (1) simplifies the task of determining random vector entropy, since it doesn't require any distribution type definition of the multidimensional stochastic variable $\boldsymbol{Y}$, which is practically unfeasible in real problems because of limited experimental data samples.

3   Entropy as a Diagnostic Model of Multidimensional Stochastic System

The essence of entropy approach in medicine is the representation of human as a complex system, which is characterized by a multidimensional random variable $\boldsymbol{Y} = (Y_1, \ldots, Y_m)$, where each Y_i element is an indicator for one of the organism subsystems, such as a risk factor. This allows us to consider the human health problem in the perspective of system analysis. Meanwhile, each risk factor is analyzed comprehensively. On one hand, it is examined separately, on the other hand – in relation to the other factors.

According to (1), the parameters of the entropy model are:

- σ_{Y_i} standard deviations of Y_i components,
- entropy indicators κ_i of distributions, $i = 1, \ldots, m$,
- coefficient of determination $R^2_{Y_k/Y_1\ldots Y_{k-1}}$ of regression dependencies of the components of random vector $\boldsymbol{Y}$, $k = 2, \ldots, m$.

The diagnostic model should explain the changes that occur in the studied object during functioning, in dynamics. Let us now consider the entropy of a random vector from this point (1).

Let the stochastic system be presented as a random vector $\boldsymbol{Y}$. Then, based on the model (1) it is possible to monitor the state of the stochastic system by analyzing the change in its entropy. This can be done as follows. Let us assume that two random vectors $\boldsymbol{Y}^{(1)} = (Y_1^{(1)}, \ldots, Y_m^{(1)})$ and $\boldsymbol{Y}^{(2)} = (Y_1^{(2)}, \ldots, Y_m^{(2)})$ describe the previous and current periods of the system functioning. We assume that the dispersions of all the random vector components are finite.

To diagnose the state of a multidimensional stochastic system we will adhere to the following steps [15]:

- system behavior definition (stable/unstable), search for time dependencies of system behavior, and critical values;
- detecting changes in the nature of the system ("randomness" or "self-organization") in critical periods;
- analysis based on the detected reason: which system element turned to be the cause of the change in its state;
- drawing conclusions about the impact of changes on the system, considering the identified critical points and their causes.

Case 1. First, consider the case when distributions of all relevant components of $Y_i^{(1)}$ and $Y_i^{(2)}$ ($i = 1, \ldots, m$) are described by the same type of distribution with some location and scale parameters. This means that $\forall i \; \kappa_i^{(1)} = \kappa_i^{(2)}$. Then the difference of entropies $\Delta H(\boldsymbol{Y}) = H(\boldsymbol{Y}^{(2)}) - H(\boldsymbol{Y}^{(1)})$ is defined as:

$$\Delta H(\boldsymbol{Y}) = \sum_{i=1}^{m} \ln \frac{\sigma_{Y_i^{(2)}}}{\sigma_{Y_i^{(1)}}} + \frac{1}{2} \sum_{k=2}^{m} \ln \frac{1 - R^2_{Y_k^{(2)}/Y_1^{(2)}\ldots Y_{k-1}^{(2)}}}{1 - R^2_{Y_k^{(1)}/Y_1^{(1)}\ldots Y_{k-1}^{(1)}}}. \qquad (2)$$

Having a pair of components $Y_i^{(1)}$ and $Y_i^{(2)}$ with the same types, instead of trialism we will have a particular case of dualism, i.e. the entropy can only change either due to the change of random vector components variance or due to the change of the correlation tightness between these components.

The contribution of any l-th component in the change of the entropy of randomness:

$$\Delta H(\boldsymbol{Y})_{V,l} = \ln \frac{\sigma_{Y_l^{(2)}}}{\sigma_{Y_l^{(1)}}}, l = 1, \ldots, m.$$

Since $R^2_{Y_m/Y_1\ldots Y_{m-1}} \geq R^2_{Y_m/Y_1\ldots Y_{m-2}} \geq \ldots \geq R^2_{Y_m/Y_1}$, the ultimate coefficient of determination $R^2_{Y_m/Y_1\ldots Y_{m-1}}$ most accurately describes the dependence of the $\boldsymbol{Y}_m$ component from the rest $(m-1)$ components. Therefore, it is expedient to assess the contribution of the l-th component in the change of the self-organization entropy through the ultimate values of determination coefficients.

$$\Delta H(\boldsymbol{Y})_{R,l} = \frac{1}{2} \ln \frac{1 - R^2_{Y_l^{(2)}/Y_1^{(2)}\ldots Y_{l-1}^{(2)}Y_{l+1}^{(2)}\ldots Y_m^{(2)}}}{1 - R^2_{Y_l^{(1)}/Y_1^{(1)}\ldots Y_{l-1}^{(1)}Y_{l+1}^{(1)}\ldots Y_m^{(1)}}}, l = 1, \ldots, m.$$

The total contribution of the l-th component in the change of random vector entropy is defined as $\Delta H(\boldsymbol{Y})_l = \Delta H(\boldsymbol{Y})_{V,l} + \Delta H(\boldsymbol{Y})_{R,l}$.

Case 2. Consider the general case when at least one pair of components $Y_i^{(1)}$, $Y_i^{(2)}$ is not described by the same type of distributions. Then the difference of entropies $\Delta H(\boldsymbol{Y}) = H(\boldsymbol{Y}^{(2)}) - H(\boldsymbol{Y}^{(1)})$ is defined as:

$$\Delta H(\boldsymbol{Y}) = \sum_{i=1}^{m} \ln \frac{\sigma_{Y_i^{(2)}}}{\sigma_{Y_i^{(1)}}} + \sum_{i=1}^{m} (\kappa_i^{(2)} - \kappa_i^{(1)}) + \frac{1}{2} \sum_{k=2}^{m} \ln \frac{1 - R^2_{Y_k^{(2)}/Y_1^{(2)}\ldots Y_{k-1}^{(2)}}}{1 - R^2_{Y_k^{(1)}/Y_1^{(1)}\ldots Y_{k-1}^{(1)}}}.$$

$$(3)$$

Since in this case $\kappa_i^{(2)} - \kappa_i^{(1)} \neq 0$, there is now third factor of the entropy change caused by the distribution type change.

Thus, the case of distribution types conservation of the random vector components is easier to implement and does not require the definition of components' entropy indicators. But this condition violation may cause significant errors in the entropy dynamics assessment, and, consequently, a decrease in the reliability of the system state diagnosis.

By monitoring the change of the $\Delta H(\boldsymbol{Y})$ entropy as a whole and of its components, we can infer the state of the studied stochastic system and detect emerging trends in its changes. Analysis of changes in each component of the random vector $\boldsymbol{Y}$ is going to reveal those of them having the greatest impact on change in entropy, and hence on change in the multidimensional stochastic system state.

4   The Entropy Model Construction on Experimental Data

The main problems of the model (1) use are involved with the estimation of the determination coefficients and the entropy indicators. Let us consider them separately.

When assessing the determination coefficient, the corresponding regression types are not known, so it is necessary to use non-parametric methods.

The determination coefficient measures the proportion of variance in the dependent variable $\boldsymbol{Y}$, that is accounted for by the variances of the factor variables $X_1, X_2, \ldots, X_m$, included in the non-linear regression model [3]:

$$R^2_{Y/X_1\ldots X_m} = \frac{\sum_{i=1}^{n} (\hat{y}_i - \bar{y})^2}{\sum_{i=1}^{n} (y_i - \bar{y})^2},$$

$$(4)$$

where $\bar{y}$ is the average value of the dependent variable, $\hat{y}_i$ – regression values, y_i are the actual values of $\boldsymbol{Y}$ variable.

There are two main approaches to the construction of nonparametric regression model – grouping and data smoothing.

In multidimensional case such known data grouping algorithms [17] as FOREL, k-means algorithm, have several disadvantages: (1) poor applicability of the algorithm under poor separability of sample into clusters; (2) the need for a priori knowledge of the cluster radius; (3) the algorithm instability (the clustering result strongly depends on the choice of the initial object).

Known smoothing methods [4] (kernel smoothing, k-nearest neighbor estimation etc.) are faced with the problem of the average neighborhood size choice. The underestimated size of neighborhood doesn't smooth the non-parametric regression sufficiently which causes an increase of the determination coefficient.

An overestimated size of neighborhood, on the contrary, causes excessive smoothing, and the determination coefficient decreases. No universal criteria have been proposed for the optimal neighborhood size selection.

Hereby, we are going to describe an algorithm to eliminate this gap. It consists in following. Assume, that we have a multidimensional sample $(x_{i1}, \ldots, x_{im}, y_i)$, $i = 1, \ldots, n$. We will form the distance matrix D, whose elements are

$$d_{ij} = \sqrt{\sum_{k=1}^{m}(x_{ik} - x_{jk})^2}, \quad i, j = 1, \ldots, n.$$

For each i-th observation we select an optimal sample of Li nearest neighborhood according to the **D** matrix $N(L_i) = \{i, i_1, \ldots, i_{L_i-1}\}$ so that the constructed equation of linear regression

$$\tilde{y}_l = a_{i0} + \sum_{k=1}^{m} a_{ik}x_{lk}, \, l \in N(L_i), \tag{5}$$

has minimal variance of the regression error

$$s_{L_i}^2 = \frac{\sum_{l \in N(L_i)}(y_l - \tilde{y}_l)^2}{L_i - m - 1} \longrightarrow \min_{3(m+1) < L_i \leq n}.$$

For optimal local sample of L_i nearest neighbors we form the value of nonparametric regression $\hat{y}_i = a_{i0} + \sum_{k=1}^{m} a_{ik}x_{ik}$, based on (5).

In this way we form a set of values $\hat{y}_i$, $i = 1, \ldots, n$, with (4) we are going to find the estimation of the determination coefficient for multiple regression.

It is obvious that, in general case, the optimal dimensions of the local samples L_i will differ. This allows us to consider the changes in the gradient of the theoretical regression function depending on the values of factor variables and the variance of the random variable. Since there were no assumptions made about the form of the regression function, the described algorithm is nonparametric.

Note that the regression coefficients estimation in (5) is done differently depending on the characteristics of the source data, for example, the least squares method [2–4,17], the least modules method [1,14], as well as robust methods [5] can be used.

The entropy model of dynamics (2) can be applied to data with the same distribution type. That is, changes in the system don't cause a change in the distribution type of the data.

For practical use of (2), it is necessary to check the corresponding $Y_i^{(1)}$, $Y_i^{(2)}$ ($i = 1, \ldots, m$) components in the uniformity of $F_i^{(1)}(x)$, $F_i^{(2)}(x)$ empirical distribution functions. This can be done in several different ways. For instance, by transforming the second sample, so that its scale and location are equal to the corresponding parameters of the first sample, then by testing the statistical hypothesis of samples H_0: $F_i^{(1)}(x) = F_i^{(2)}(x)$ homogeneity. When estimating the location parameter (if it exists) we are going to use the mathematical expectation, and when estimating the scale parameter – the standard deviation.

There are many nonparametric criteria developed to test the statistical hypothesis of homogeneity, the most common are two-sample Smirnov test and omega-square (Lehmann-Rosenblatt) [9]. In case of the statistical hypothesis rejection we are going to use model (3) instead of (2). In this case we have to estimate the entropy indicators.

The estimation of the random variable entropy indicator comes to the normalization of the random variable (divided by the standard deviation) and to the definition of the differential entropy of the normalized random variable.

Assume, we have some sample of the random variable $\mathbf{X}$, of size n: x_i, $i = 1, \ldots, n$ and of a unit variance. If its sample variance $s_x^2 \neq 1$, then we will divide all the x_i observations by s_x. The problem of determining the entropy indicator of the random variable $\mathbf{X}$ comes to the differential entropy estimation, because in this case

$$\kappa = H(X) = -\int_{-\infty}^{+\infty} p_X(x) \ln p_X(x) dx.$$

Note that currently there is a number of algorithms suggested for solving this problem. However, in general, those special algorithms use some a priori information about the properties of the random variable $\mathbf{X}$. Let us describe an algorithm [15], which requires no prior knowledge of the random variable $\mathbf{X}$. It is based on the following formula:

$$\hat{\kappa} = \hat{H}(X) = -\sum_{j=1}^{L} \hat{p}_j \ln \hat{p}_j + \ln h, \tag{6}$$

where $\hat{p}_j = \frac{n_i}{n}$ is the probability estimation of falling into j-th interval, L – number of intervals with length h, continuously covering the range of experimental data x_i, $i = 1, \ldots, n$; n_j – the number of observations fell into a j-th interval, n – the sample size.

The term $\ln h$ in (6) was introduced to eliminate the offset from the formula of information entropy. The main problem of using (6) is the selection of the L intervals number, which depends on the distribution of random variable $\mathbf{X}$. At too small L intervals number, the $\hat{H}(X)$ estimate will be overrated and at too many intervals – understated. Figure 1 is a schematic graph illustrating the dependence of entropy estimation and the number of intervals.

Using the method of statistical tests for thirty distribution types the empirical formula for the optimal intervals number was obtained

$$\hat{L} = 1.072 n^{0.968 - 0.231 I_{0.8}} - 2.098 \gamma_1 - 1.789, \tag{7}$$

where $I_{0.8} = x_{0.9} - x_{0.1}$ is the interdecile range, $x_{0.1}$ and $x_{0.9}$ are quantiles of the empirical distribution $\hat{F}_n(x)$ for levels 0.1 and 0.9, correspondingly.

Equation (7) also accurately describes the dependence of the intervals number on the sample size, the asymmetry coefficient and interdecile range.

Some of the statistical test results (tests number equals to 50,000) are shown in Table 1. Here are the 95% confidence intervals of the entropy indicators estimates according to (6) and (7).

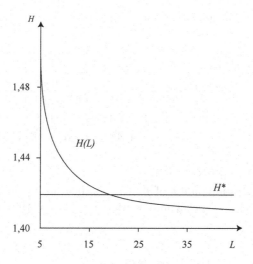

Fig. 1. Entropy estimate dependence on L number of intervals, H^* – theoretical value of the entropy

Table 1. 95% confidence intervals of the estimates of $\hat{\kappa}$ entropy indicators according to (6) and (7)

Distribution $F_X(x)$	κ	$\hat{\kappa}$		
		$n = 250$	$n = 500$	$n = 1000$
Weibull	1	(1.000; 1.026)	(0.990; 1.008)	(0.989; 1.003)
Hamble	1.329	(1.349; 1.365)	(1.337; 1.347)	(1.332; 1.338)
Gamma-distribution	1	(1.000; 1.026)	(0.990; 1.009)	(0.989; 1.002)
Laplace	1.347	(1.349; 1.365)	(1.346; 1.358)	(1.346; 1.353)
Logistic	1.405	(1.402; 1.411)	(1.403; 1.409)	(1.404; 1.408)
Log-normal	0.649	(0.639; 0.717)	(0.632; 0.693)	(0.633; 0.679)
Normal	1.419	(1.415; 1.423)	(1.418; 1.422)	(1.419; 1.421)
Pareto	0.779	(0.783; 0.836)	(0.762; 0.805)	(0.763; 0.795)
Uniform	1.24	(1.224; 1.235)	(1.230; 1.238)	(1.234; 1.239)
Trigonometric	1.395	(1.388; 1.395)	(1.389; 1.394)	(1.391; 1.394)
Exponential	1	(1.001; 1.027)	(0.990; 1.009)	(0.990; 1.003)

The studies have shown that the estimates are not resistant to outliers. Table 2 shows the 95% confidence intervals of the entropy indicators estimates for various distributions with the samples where three largest values were increased twice. We can see that the estimation results are significantly offset compared to the theoretical values of κ. We could also note that with the increase of the sample size n the offset decreases. Similar results are obtained for practically all distributions.

Table 2. 95% confidence intervals of the $\hat{\kappa}$ estimates of κ entropy indicators according to (6) and (7), having outliers

Distribution $F_X(x)$	κ	$\hat{\kappa}$		
		$n = 250$	$n = 500$	$n = 1000$
Logistic	1.405	(1.293; 1.313)	(1.329; 1.341)	(1.356; 1.364)
Normal	1.419	(1.347; 1.361)	(1.374; 1.382)	(1.394; 1.398)
Trigonometric	1.395	(1.188; 1.200)	(1.287; 1.293)	(1.344; 1.348)

To ensure the sustainability of entropy indicators estimation we can use known procedures of censoring and winsorising [11] of the initial sample x_i, $i = 1, \ldots, n$.

5 The Experimental Results

Let us investigate the opportunities of the stochastic system entropy modeling on the example of population's biological risk factors analysis in terms of chronic non-communicable diseases (NCD) prevention.

The peculiarity of biological risk factors is that they can be considered as functioning parameters of the human organism homeostatic systems. They become NCD risk factors when their values exceed the "cut points" which are associated with the significant growth of death due to various reasons and which are determined during epidemiological studies. This significantly reduces the cases of disease diagnosis at early stages.

To analyze the population entropy we chose parameters characterizing the risk factors that have important prognostic value and can be measured objectively and quantitatively. Factors like blood pressure, overweight, high blood sugar, dyslipidemia are among them.

Thus, we have four groups of NCD risk factors: blood pressure (systolic and diastolic blood pressure), weight (body mass index), dyslipidemia (level of total cholesterol, triglycerides, low density lipoprotein cholesterol, high density lipoprotein cholesterol), blood sugar level. Because of the strong correlation between the systolic and diastolic blood pressure, and also between the levels of total cholesterol, low and high density lipoprotein cholesterol, triglycerides, for the further entropy analysis four subsystems were identified: "Total cholesterol" (TC), "Systolic blood pressure" (SBP), "Body mass index" (BMI), "Blood sugar" (BS).

It should be noted that the risk factors prevalence, as well as the average levels of their parameters, are significantly affected by age. Particularly, among the population we studied the healthy individuals over 35 years old were of a very small amount. Therefore, for analyzing the population entropy changes based on the health state two age groups of equal ranges were formed: 18–26 years, 27–35 years.

The statistical material was obtained as a result of a complex, in-depth clinical and epidemiological survey among male rural population with cluster sampling. For all the patients the required complex of clinical, laboratory and instrumental examinations were carried out to get qualified conclusions about the health state. The work was done by specialists of the Department of Hospital Therapy and Family Medicine of the Chelyabinsk State Medical Academy and the Chelyabinsk Regional Clinical Hospital No. 1.

The assessment and analysis of the health state was carried out according to the detailed group classification proposed in [8], on criteria "healthy", "practically healthy", and "diseased":

- Healthy: (1) during the examination period did not suffer from acute and chronic diseases; no anomalies of health were revealed during the checkup. (2) rarely suffered from acute respiratory or other acute diseases (no more than 3 times during a year); no anomalies of health were revealed during the checkup.
- Almost healthy: (1) those having such risk factors as overweight, hypercholesterolemia (HCL) (with the absence of diseases and if the checkup didn't reveal any health anomalies); (2) those who often (more than 4 times a year) suffered from acute diseases and no anomalies of health were revealed during the checkup. This group also includes those who at the same time have such risk factors as smoking and physical inactivity; (3) those, for whom during extra examinations there were various functional anomalies or symptoms without diagnosis revealed: harbingers of disease, states after an illness or injury; (4) patients with diseases that are characterized to lose their significance when reaching certain age, with long intervals between attacks.
- Diseased: (1) patients who have chronic diseases of various compensation levels; (2) people with disabilities and long-term consequences of disease and injury, characterized by disability or deterioration in life quality.

HCL is diagnosed at the level of total cholesterol (TC) not less than 5.17 mmol/L, Overweight is diagnosed at the level of BMI 25–29.9, which is determined by formula weight(kg)/(height(m))^2.

The study objective is the investigation of the changes in parameters entropy, characterizing the biological risk factors for NCD for men in different health groups.

In [16] a population health analysis was done using entropy modeling, but the presence of outliers in samples was ignored and an assumption was made that each population risk factor preserves its distribution while health state changes. These assumptions were not sufficiently reasonable in this case. Therefore, we are going to do more detailed analysis.

The entropy analysis of the parameters, characterizing the main risk factors of NCD, was held within 420 people.

The results of population health entropy analysis for groups "healthy", "almost healthy" and "diseased" are presented in Tables 3 and 4.

With the deterioration of population health state an increase occurs in the total and randomness entropy for all the risk factors. This can be explained

Table 3. The entropy levels for groups of "healthy", "almost healthy" and "diseased" people

Age, years	Health state	Randomness entropy	Self-organization entropy	Total entropy
18–26	Healthy	5.500	−0.514	4.986
	Almost healthy	7.131	−0.578	6.553
	Diseased	7.847	−0.696	7.151
27–35	Healthy	5.731	−0.299	5.432
	Almost healthy	8.376	−0.542	7.834
	Diseased	8.720	−0.781	7.939

Table 4. The dynamics of the entropy at deterioration of health state

Age, years	Health state	Dynamics of Randomness entropy	Dynamics of Self-organization entropy	Dynamics of Total entropy
18–26	"Healthy" → "Almost healthy"	1.631	−0.064	1.567
	"Almost healthy" → "Diseased"	0.717	−0.118	0.598
	"Healthy" → "Diseased"	2.348	−0.182	2.165
27–35	"Healthy" → "Almost healthy"	2.644	−0.243	2.401
	"Almost healthy" → "Diseased"	0.345	−0.240	0.105
	"Healthy" → "Diseased"	2.989	−0.483	2.506

by the fact that additional damaging effects of NCD are being added to the pathological influence of risk factors on the human organism individually and to the whole population in general.

This conclusion has of practical value, because, if one of the studied populations has higher entropy, without holding expensive and long examinations, we can make a preliminary conclusion about the worst health state of this group and select it for further examination.

The self-organization entropy, on the contrary, with the deterioration of population health, is reduced which means stronger relationships between the subsystems. This can be explained by the idea, that the disease development within the organism doesn't happen isolated. On the other hand, with the disease development some subsystems may adapt to others, compensating the defects in functioning, i.e. some substitution effect can be noted. The less expressed downward trend in self-organization entropy for the younger group (18–26 years) is likely to occur due to the fact that the final growth and formation of the human body

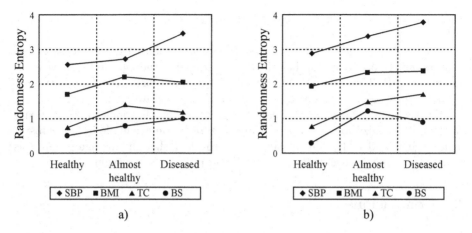

Fig. 2. The randomness entropy for each risk factor for the age groups: (a) 18–26 years; (b) 27–35 years

happens before 25 years, so for young people there is a greater scatter of biological indicators due to the growth effect.

For a more detailed analysis of the subsystems individually we have listed the calculation results of randomness entropy for the subsystems TC, SBP, BMI, BS (Fig. 2) and the results of determination coefficient for each risk factor the regression to the rest of the factors (Fig. 3).

On Fig. 2 we can see that the randomness entropy is growing by about the same rate for each factor but on the transition from "Almost healthy" to "Diseased" in some cases a slight decrease is observed (it is less expressed than

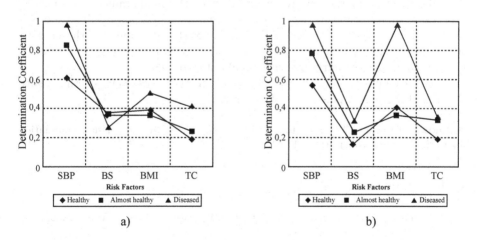

Fig. 3. Determination coefficients for each risk factor regression to the rest factors for age groups: (a) 18–26 years; (b) 27–35 years

the growth of randomness entropy for those factors on the transition from state "Healthy" to "Almost healthy").

On Fig. 3 we can see that the greatest contribution to the decrease of the self-organization entropy at health deterioration for the age of 18–26 years is made by SBP. This means that the blood pressure starts correlating strongly with the other risk factors at deterioration of health state.

For the age group of 27–35 years the greatest contributions to the decrease of the self-organization entropy at health deterioration are made by two subsystems – SBP and BMI. This speaks about the fact that in this age group it is necessary to pay a special attention to blood pressure and weight.

6   Conclusions

The article examined the possibilities of using the entropy modeling in medical diagnostics. As the study objective, the analysis for the change of the parameters entropy characterizing the biological risk factors for chronic non-communicable diseases for men from different health groups was carried out.

The entropy analysis application in the prevention of chronic non-communicable diseases, combined with the population health assessment, allows to determine the most critical changes in the population health and find out which subsystem of the human organism affected that change. The results proved the adequacy of the constructed entropy model. It is of interest the entropy modeling development for health state changes assessment of individual patients.

The proposed approach has limitations, advantages and shortcomings. The restriction is the adequacy requirement of the multidimensional stochastic system representation being studied as a random vector.

The approach disadvantages are the sensitivity to abnormal emissions in the initial data and the relative complexity of the nonparametric smoothing procedure while calculating determination coefficients. The approach advantages include the no need of restoring the random vector distribution, the entropy representation in the form of a diagnostic model, the ability to work with discrete random components, and work with small data samples.

The development of entropy modeling for the assessment of the health changes of individual patients is of interest.

Acknowledgments. The reported study was supported by the Russian Foundation for Basic Research (project No. 17-01-00315).

References

1. Bloomfield, P., Steiger, W.: Least Absolute Deviations. Theory Applications, and Algorithms. Brikhauser, Basel (1983). https://doi.org/10.1007/978-1-4684-8574-5
2. Chavent, G.: Nonlinear Least Squares for Inverse Problems. Springer, Heidelberg (1993)
3. Greene, W.: Econometric Analysis. Prentice Hall, Upper Saddle River (2011)

4. Hardle, W.: Prikladnaya neparametricheskaya regressiya. [Applied nonparametric regression]. Mir, Moscow (1993). (in Russian)
5. Huber, P.: Robust Statistics. Wiley, Hoboken (1981)
6. Kiryanov, B., Tokmachev, M.: Matematicheskie modeli v zdravookhranenii. [Mathematical models in health care]. NovSU, Veliky Novgorod (2009). (in Russian)
7. Klimontovich, J.: Vvedenie v fiziku otkrytykh sistem. [Introduction to the physics of open systems]. Yanus-K, Moscow (2002). (in Russian)
8. Marchenko, A.: Kriterii otsenki sostoyaniya zdorov'ya naseleniya pri ego kompleksnom izuchenii. [Criteria for population health assessment in complex study]. Sov. Health Care (2), 23–28 (1979). (in Russian)
9. Orlov, A.: O proverke odnorodnosti dvukh nezavisimykh vyborok. [Verification of the homogeneity of two independent samples]. Ind. Lab. **69**(1), 55–60 (2003). (in Russian)
10. Shannon, C.: Mathematical theory of communication. Bell Syst. Tech. J. **27**(379–423), 623–656 (1948)
11. Smolyak, S., Titarenko, B.: Ustoychivye metody otsenivaniya. [Robust estimation methods]. Statistics, Moscow (1980). (in Russian)
12. Timashev, S., Tyrsin, A.: Entropy approach to risk-analysis of critical infrastructures systems. In: Vulnerability, Uncertainty, and Risk. Analysis, Modeling, and Management: Proceedings of the ICVRAM 2011 and ISUMA 2011 Conferences, pp. 147–154 (2011). https://doi.org/10.1061/41170(400)18
13. Tsinker, M., Kiryakov, D., Kamaltdinov, M.: Primenenie kompleksnogo indeksa narusheniya zdorov'ya naseleniya dlya otsenki populyatsionnogo zdorov'ya v permskom krae. [Application of complex index of population health disorders to estimate the population health in perm region]. Proc. Samara Sci. Center Russ. Acad. Sci. **15**(3), 1988–1992 (2013). (in Russian)
14. Tyrsin, A.: Robust construction of regression models based on the generalized least absolute deviations method. J. Math. Sci. **139**, 6634–6642 (2006). https://doi.org/10.1007/s10958-006-0380-7
15. Tyrsin, A.: Entropiynoe modelirovanie mnogomernykh stokhasticheskikh sistem. [Entropy modeling of multidimensional stochastic systems]. Nauchnaya Kniga, Voronezh (2016). (in Russian)
16. Tyrsin, A., Kalev, O., Yashin, D., Lebedeva, O.: Otsenka sostoyaniya zdorov'ya populyatsii na osnove entropiynogo modelirovaniya. [Assessment of population health based on entropy modeling] (2015). http://www.matbio.org/article.php?journ_id=19&id=234. (in Russian)
17. Zaguroyko, N.: Prikladnye metody analiza dannykh i znaniy. [Applied methods of data and knowledge analysis]. Publishing House of the Institute of Mathematics, Novosibirsk (1999). (in Russian)

Students Learning Results Prediction with Usage of Mixed Diagnostic Tests and 2-Simplex Prism

Anna Yankovskaya[1,2,3,4(✉)], Yury Dementyev[4], Artem Yamshanov[3], and Danil Lyapunov[3,4]

[1] Tomsk State University of Architecture and Building, Tomsk, Russia
`ayyankov@gmail.com`
[2] National Research Tomsk State University, Tomsk, Russia
[3] Tomsk State University of Control Systems and Radioelectronics, Tomsk, Russia
[4] National Research Tomsk Polytechnic University, Tomsk, Russia

Abstract. Students learning results prediction is one of the "hottest" problems in modern learning process. We describe mathematical framework of students learning results assessment on the basis of mixed diagnostic tests (MDT). This framework includes cognitive graphic tools 2-simplex and 2-simplex prism, being the powerful means of data visualization for learning outcomes evaluation and efficient goal-setting. A new approach to the prediction of students' learning results based on MDT and cognitive graphic tools is proposed. Students test results for the e-learning course "Selected Chapters of Electronics" and examples of their learning outcomes cognitive visualization are given. The proposed approach to prediction and cognitive visualization is discussed. Cross-platform software implementation specificity of cognitive graphic tools invariant to problem areas is described.

Keywords: Learning · e-learning · Prediction · Mixed diagnostic test · Cognitive tool · Data visualization · 2-simplex prism

1 Introduction

The development of learning intelligent systems based on tests is an urgent problem [1–4] today. New information technologies provide a number of innovative and very promising technologies for teaching and learning. A substantial group of these technologies is related to the online education or to the blended learning involving limited interaction between students and educators [5,6].

Students learning results prediction is one of the "hottest" problems in modern learning process. In the paper [7] the analysis of interdependence between students learning styles, their learning behavior, and academic success is performed. Students' age, experience, and academic performance type are taken into account. Modern educational process is characterized by sufficient changes in students' activity fields which are entirely outlined in [8]. Taking into account

© Springer Nature Switzerland AG 2019
V. V. Strijov et al. (Eds.): IDP 2016, CCIS 794, pp. 106–121, 2019.
https://doi.org/10.1007/978-3-030-35400-8_8

each student's personal qualities adds a great value to the process of learning. Such qualities include subject understanding level, residual knowledge and skills, and, in particular, student's interest in the subject.

Prediction of student learning outcomes is a relevant problem. Modern world's trends in e-learning technologies are represented in the survey [9] and in the monograph [10]. The models of learning process and teaching scientific disciplines are described on the example of mathematics in [11]. A necessary component for predicting student learning outcomes is monitoring students' activities. Learning intelligent systems application is used for individual learning plans formation in a number of leading universities on the basis of data collected while input testing [12]. The impact of e-learning peculiarities on students' motivation has been revealed in research conducted by a number of scientists [13]. The research of human-computer interaction within the framework of e-learning allowed to create a predictive model for the advanced electronic courses development, one of which is described in [14]. It is crucial to develop user-oriented interface design taking into account user's individual abilities. One of such interfaces is described in [15].

We note that the concept of gamification in modern education can have a positive impact on the learning process, which is expressed in greater student satisfaction, motivation, and increased engagement in the learning process [16]. This concept needs to be developed for integration into intelligent learning and testing systems, supplemented by the components predicting student's learning outcomes.

A model of students learning and testing using mixed diagnostic tests, which are a compromise between unconditional and conditional components, was initially proposed in [17–26]. For the first time development of intelligent learning and testing system interface with cognitive graphic tools n-simplex ($n \in \{2, 3\}$) and 2-simplex prism was conducted by Anna Yankovskaya and co-authors and published in [26–30]. The essentials of cognitive graphic tools—2 simplex and 2-simplex prism for indicating students' learning outcomes and subsequent decision-making will be represented in this paper. Many publications on this topic were presented at the IASTED conference and published in its proceedings. Significant survey of the publications on the topic under consideration was performed by the same research group [20–26, 31–33].

In this paper we propose and describe a new approach to the prediction of learning outcomes, based on the usage of mixed diagnostic tests and cognitive graphic tools 2-simplex and 2-simplex prism. We hope that the approach will alleviate significantly the learning process both for students and for educators, allowing to find the proper learning trajectory on each subsequents step of the learning process.

2   Basic Definitions

The following concepts and definitions from the papers [3, 17, 21, 22] are used.

Student is a person participating in learning and testing.

Learning axis is axis corresponding to an independent direction of student evolution, e.g. theory knowledge level, ability to practical tasks solving, ability of electric circuit design and research.

Test result is a set of coefficients, numerically representing student's evaluation. Each coefficient corresponds to a particular learning axis.

Prediction result is a set of coefficients, numerically representing the student's assessment which he is likely to achieve in a preassigned time interval.

Object is something at which investigation is aimed, e.g. a student.

Pattern is a typical or generalized representative of some class, e.g. students with innovative mindset.

Diagnostic test [3,34] is a set of features that distinguish any pairs of objects that belong to different patterns.

A diagnostic test is called *non-redundant* (dead end [34]) if it involves a non-redundant number of features.

Unconditional component of the diagnostic test is characterized by the simultaneous presentation of all its constituent test tasks during the decision making.

Conditional component of the diagnostic test is characterized by the sequential presentation of test tasks, depending on the assessment of the previous results.

Mixed diagnostic test (MDT) [3,17] is a compromise between unconditional and conditional components. Conditional component of MDT is used to establish the sequence of test tasks based on the results of previous work. Testing of this kind is apt to show a student how to go into further study in case of irregularities or difficulties appearing in the learning process.

MDT tree is the structure to display relationship between the different elements of MDT. The root node is necessarily correlated with the unconditional component of MDT. Each of the remaining nodes is associated with either unconditional or conditional component of MDT. Tree edges is a set between the nodes, i.e. between the different components of MDT.

3   Framework of Students' Learning Outcomes Evaluation Based on MDT

The basis of MDT construction is fully described in the papers [21–23]. This paragraph describes only part necessary for understanding the article and changes that were made for this research.

An educational course is divided into some sections; each of those consists of didactic units. Material for all didactic units should be covered by an educational test. Educational test is a set of test tasks which assess student knowledge level. The first part of educational test is unconditional component of the MDT, i.e. test tasks that are introduced to students at random sequence. An example of the MDT search tree for the course "Informatics" [22] is given in Fig. 1. Root node of the search tree is associated with unconditional component of educational test (Sects. 1, 2 and 3). Solving test tasks in the second part (conditional component test) depends on what test task was previously presented and solved on an appropriate level of the MDT search tree. Each branch of the tree represents an admissible sequence of test tasks to select the section that leads to a leaf. It should be noted, that an order of test tasks at the same level of the search tree does not matter. A test task can be one of the following tasks: a closed form with the only answer, a closed form with multiple answers, a matching test, a sequence test etc.

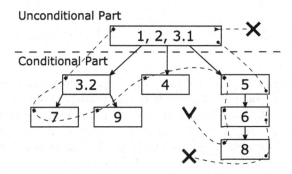

Fig. 1. Example of MDT search tree

Each test task is mapped to a set of coefficients for each learning axis. For each of the tasks included in the test weights matrix and the questions weight, which student will get if the answer is correct, should be formed (Table 1). The rows of weights matrix are associated with the test tasks given to the students, where n—the number of all possible questions (the strength of the test). The columns of weights matrix are associated with coefficients for each learning axis. In the current research the number of learning axes is 3: theory knowledge level, ability to solve practical tasks, ability of electric circuit design and research. Student's test result is a set of coefficients sum along all the axes for all correctly solved tasks. If a student scores less than 50% correct answers for any test node, testing will be failed. Dashed lines in Fig. 1 are used to represent admissible trajectory of learning or testing process, the first is successful (marked with symbol ✓), i.e. all test nodes are complete, the second and the third are failed (marked with symbol ✗), i.e. for the second the student scored less than 50% of unconditional part, and for the third the student scored less than 50% for test node 8 in conditional part.

4   Family of 2-Simplex Cognitive Graphic Tools

For representing the students' learning outcomes with regard to the learning axes, it is reasonable to use cognitive graphic tool based on 2-simplex [26]. For this purpose we need to convert test results (i.e. feature space) into the students' proximity patterns corresponding to the direction of student evolution (i.e. pattern space).

Table 1. Learning axes weight matrix

	Theory (1)	Problems solving (2)	Electric circuit design (3)
1	$w_{1,1}$	$w_{1,2}$	$w_{1,3}$
2	$w_{2,1}$	$w_{2,2}$	$w_{2,3}$
...	...	...	...
i	$w_{i,1}$	$w_{i,2}$	$w_{i,3}$
...	...	...	...
n	$w_{n,1}$	$w_{n,2}$	$w_{n,3}$

For the first time, the transformation of features space into patterns space based on the logical-combinatorial methods and properties of n-simplex are suggested in the paper [35]. System of visualization TRIANG for decision-making and its justifications using cognitive graphic tools [36] is constructed on the basis of theorem proved by Yankovskaya [35]. This theorem was used in more than 30 applied intelligent systems and in three intelligent instrumental software (frameworks for creating applied intelligent systems) revealing different kinds of regularities and decision-making of the following types: diagnostic, organization and control, classification; and their justifications. Some parts of the chapter were already published in a set of papers, for example [30].

Theorem. Suppose $a_1, a_2, \ldots, a_{n+1}$ is a set of simultaneous non-zero numbers where n is the dimension of a regular simplex. Then, there is one and only one point at which that following condition $h_1 : h_2 : \ldots : h_{n+1} = a_1 : a_2 : \ldots : a_{n+1}$ is correct, where $h_i, (i = 1, 2, \ldots, n + 1)$ is the distance from this point to the i-th side [35].

Coefficient $h_i, (i = 1, 2, \ldots, n + 1)$ represents the degree of conditional proximity of the object under study to the i-th pattern. The advantage of this fact is that the n-simplex possesses the constancy property of distances (h) sum from any point to each side and the property of ratios preservation $h_1 : h_2 : \ldots : h_{n+1} = a_1 : a_2 : \ldots : a_{n+1}$.

The main function of n-simplex is a disposition representation of the object under study among other objects of a learning sample [36]. Additionally, n-simplex has other useful functions for a decision-making person. One of these functions is a representation of some numerical values, e.g. prediction confidence region, which will be described below. Last step required to visualize the transformation from n-simplex space into Cartesian 2D reference frame [28,38,39]. An example of data visualization using 2-simplex cognitive graphic tool is given in Fig. 2.

Sides of the 2-simplex (triangle edges) are associated with patterns (classes), circles with big radius are objects under study and circles with small radius are learning sample objects. Each pattern corresponds to a color which can be preassigned by an expert. For example, red, yellow, and green colors are used. The distance from an object to a side is inversely proportional to the object proximity to the pattern corresponding to the side. The distance for the object

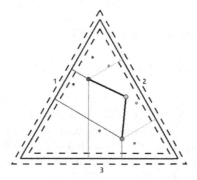

Fig. 2. Data visualization using 2-simplex (Color figure online)

under study is displayed as colored perpendicular lines to 2-simplex sides. Color of each line corresponds to the pattern color. The color of object under study or objects from a learning sample is mapped to the pattern, which is revealed for a specific object. An object color is mapped with an associated pattern (the nearest pattern or pattern determined by an expert).

Visualization of confidence region for prediction in 2-simplex is not a trivial task that is associated with a non-standard 2-simplex space and distance estimation in 2-simplex space. The following algorithm for this visualization is proposed:

1. Three perpendiculars are drawn from the target point a (i.e., visualization of object under investigation) to each side.
2. Three points b, c, d are drawn on each perpendicular on the same distance s from the point a. The distance s is the size of predefined confidence region. Calculation algorithm of the distance s can be configured by an expert.
3. Three lines are drawn through points b, c, d parallel to the sides of 2-simplex.
4. A triangle formed by these lines is the confidence region for prediction.

Such visualization is equivalent to the visualization of confidence region for prediction in a square shape in Cartesian 2D reference frame. An example of described calculations is given in Fig. 3.

For visualization of dynamic processes it is reasonable to use cognitive graphic tool "2-simplex prism" (Fig. 4) which is based on 2-simplex and represents the equilateral triangular prism which contains 2-simplexes in bases and cross-sections which correspond to the fixed time moments. This is necessary for a big amount of problems and such cross-disciplinary areas as medicine, economics, genetics, building, radioelectronics, sociology, education, psychology, geology, design, ecology, geo-ecology, eco-bio-medicine, etc.

The distance from the prism base to the i-th 2-simplex h'_i corresponds to the fixed time moment of object's features and it is calculated from the following formula:

$$h'_i = H' \cdot \frac{T_i - T_{min}}{T_{max} - T_{min}}$$

where H' is a length of a 2-simplex prism preassigned by a user and corresponds to the learning duration, T_i is a features fixation timestamp of the object under study for the i-th examination, T_{min} is a timestamp of features fixation of the object for the 1-st examination, T_{max} is a timestamp of features fixation of the object for the last examination.

5   Prediction of Students' Learning Outcomes and Cognitive Graphic Tools for Visualization

One of the main issues of the research at the current step is fast creation of a simple prototype for the prediction model for a learning process. It is very important because it allows to solve the following problems: (1) to remove obstacles for solving other tasks of the current research; (2) to get an estimation of learning process trajectory influence and its prediction visualization given for the students on the speed and the quality of their learning; (3) to reveal how the cognitive graphic tools help students improve their learning outcomes and facilitate subsequent efficient goal-setting and integrate the proper visualization library components in e-learning platform (e.g. Moodle). As a result, the main attention at this step is focused on inputs and outputs of this model and inner details of realization are shifted to the second place. Additionally solved task is a possibility of easy replacement of one mathematical model on another in the future.

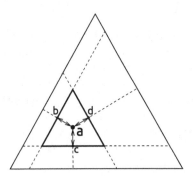

Fig. 3. Calculation of confidence region for prediction in 2-simplex

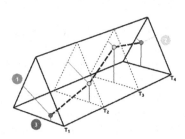

Fig. 4. An example of the 2-simplex prism

The prediction model has the following inputs: (1) history of previous student testing results, which are obtained by MDT, described before; (2) a timestamp of prediction. The output of prediction model is a predicted result which a student should get at test in the preassigned timestamp.

A knowledge estimation for each student is performed by different and independent among themselves axes (directions). The axes should be predefined by an expert. They should be orthogonal and independent, and value for the one axis should not depend on values for the others. A real data usually show some dependencies which were unknown to an expert before. Revealing of such dependencies is a separate task and this issue is beyond the scope of this paper. Assume, that orthogonal requirement is completely fulfilled and prediction for each knowledge level axis can be performed independently based on the following formulas:

$$E_{k,1} = f(k, E_{n,1}, E_{n-1,1}, \ldots, E_{n-m+1,1}),$$
$$E_{k,2} = f(k, E_{n,2}, E_{n-1,2}, \ldots, E_{n-m+1,2}),$$
$$\ldots$$
$$E_{k,l} = f(k, E_{n,l}, E_{n-1,l}, \ldots, E_{n-m+1,l}),$$

where:

- l is a number of the learning axes ($l \geq 1$, in the paper we use $l = 3$);
- n is a number of already performed tests ($n \geq 1$);
- m is a number of tests which is required for a prediction ($1 \leq m \leq n$).
- k is the number of the predicted tests ($k > n$).
- $E_{i,j}$ is student's result for the i-th test for the j-th axis ($i \leq n$ for already performed tests, $i > n$ for predicted tests);
- $f(k, E_{n,1}, \ldots, E_{n-m+1,1})$ is a prediction function;

In the current research we use a simple extrapolation polynomial function for prediction. Polynomial degree p is configurable and can be used for its influence estimation on a prediction quality. For a polynomial function the last $p + 1$ result of already performed tests by a student for each axis ($m = p + 1$) are used. A system of linear equations is constructed and solved via Gauss method. It should be noticed, that the value of k is also configurable, so prediction can be performed for any future step, not only for the next one.

Confidence region prediction is calculated as delta between predicted and real result for the last performed test. Process of this calculation for one axis is shown in Fig. 5.

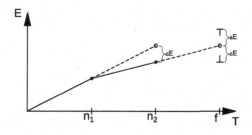

Fig. 5. Example of confidence region prediction calculation

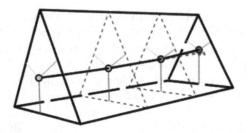

Fig. 6. 2-simplex prism application for prediction of students' learning results (Color figure online)

Applying of 2-simplex prism for "Power Electronics" discipline was firstly described in the paper [24]. Some fragments are given herein.

Students' test results were taken for checking the prediction model. They were obtained from third year students of speciality "Electrical Power Engineering". The course "Selected Chapters of Electronics" is aimed at students' knowledge, abilities, and skills development within the discipline "Power Electronics" using English language. The course is divided into 4 modules. Each module contains learning materials and tests for students' essential skills assessment in three learning axes: theory knowledge level, ability to practical tasks solving, ability of electric circuit design and research. Four tests were performed: input test, 2 intermediate tests and one pass-fail test. Each test reveals students' essential skills across the aforementioned axes.

Source data for this visualization are presented in Table 2. Example of learning trajectory with a 2-simplex prism for one student is given in Fig. 6.

All test results are represented as points within 2-simplex, which are placed in bases or cross-sections of the 2-simplex prism. Each side of a 2-simplex prism is associated with one pattern. For such technical discipline as "Power Electronics" the following learning axes are rational: level of theoretical knowledge (in figure associated with red color), skill of practical problem solving (yellow), skill of electric circuits design (green). Each axis is associated to a particular pattern. The distance from the side of 2-simplex prism to a point is inversely proportional to the revealed level of skill. Distance from the base of 2-simplex prism to a specific 2-simplex (cross-section) corresponds to the time interval between the course start and specific test (or prediction). Height of 2-simplex prism corresponds to the time interval of the whole course plus some time after finishing the course, which is used for prediction of future student results.

The polyline inside the 2-simplex prism represents evolution of a student knowledge level. For prediction quality estimation, two lines for the last step are used: solid and dashed. Solid line represents students' progress obtained from their testing. Dashed line represents predicted students' progress based on their previous tests. Confidence region of prediction is represented in the shape of a triangle formed from dashed lines.

As a prediction polynome degree is configurable, it can be used for the impact research of this parameter on prediction quality. Examples of predictions constructed with different polynome degrees are given in Fig. 7. Variation

Table 2. Source data for visualization

Stud.#	Theory				Problems solving				Electric circuits design			
	Input test	Test 1	Test 2	Pass-fail test	Input test	Test 1	Test 2	Pass-fail test	Input test	Test 1	Test 2	Pass-fail test
1	63	72	80	83	70	80	85	84	75	77	85	90
2	54	56	60	68	60	56	65	67	59	65	70	75
3	53	56	60	63	52	60	55	61	60	64	70	68
4	58	63	65	67	65	62	70	76	68	71	75	82
5	51	50	54	52	52	56	57	55	54	53	56	60
6	65	63	68	75	73	75	72	75	76	73	78	84
7	60	65	63	68	65	68	75	74	72	76	76	82
8	55	58	54	55	53	60	62	66	60	62	60	65
9	72	76	68	75	76	68	63	70	80	72	83	85
10	51	51	53	54	51	56	58	55	51	55	53	54
11	52	55	53	64	62	70	68	75	60	56	72	73
12	68	64	69	75	70	75	76	80	75	68	74	80
13	52	55	54	60	51	52	60	60	60	65	72	68
14	56	58	54	56	75	79	82	84	68	75	81	75
15	82	86	81	86	75	76	72	70	75	78	69	70
16	64	67	67	72	55	54	60	65	52	58	70	68
17	51	52	51	52	55	62	61	60	54	66	68	72
18	60	65	68	67	56	58	52	54	62	60	64	62
19	51	52	51	52	51	51	53	52	51	55	60	65
20	65	67	65	68	60	65	62	66	72	76	78	80
21	60	62	66	65	56	65	72	75	52	55	56	59

of prediction polynome degree shows that for a majority of students prediction results obtained via using linear polynome (Fig. 7a) are better than prediction results obtained via using square polynome (Fig. 7b). On this basis the following hypothesis can be formulated: the second derivative is not constant for a learning process and cannot be used for correct prediction. But this hypothesis confirmation requires a full-scale research with the analysis of large amount of tests results data. Also a huge experiment base can be used for the application research of other polynomes or even other mathematical prediction models.

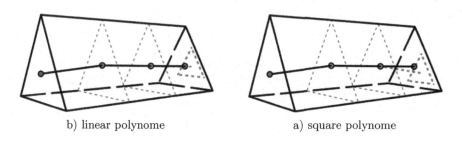

b) linear polynome   a) square polynome

Fig. 7. Polynome degree influence on prediction quality

In most cases the chosen simple prediction model shows appropriate quality of prediction. That is why it is going to be used in the first version of an intelligent learning technology development.

There are good grounds to believe that proposed approach allows to obtain the following results:

1. Illustrative representation of a student learning trajectory and trajectories comparison for different students.
2. Revealing and representation of a student learning speed.
3. Revealing of incorrect test results, e.g. revealing cheating on a test.
4. Predicting and modeling a student learning process.
5. Increasing students' motivation to a high-performance learning by increasing in-group competition, which is accessible through providing both comparison of one student's results and the other students' results.

It should be noticed, that all figures are rendered with the use of visualization library, which is currently at the early stage of active development. So they may contain some small visualization problems, e.g. incorrect lines overlapping, and sometimes necessary objects do not render at all, e.g. lettering for sides. Interactive demonstration for cognitive graphic tool, described in the present paper, is available by URL [37].

6   Specificity of Software Implementation of Cognitive Graphic Tools and Their Integration

Firstly, prototype of cognitive graphic tools (CGT) visualization library was implemented as C# [38] class library. But this implementation has technical restrictions with integration into other intelligent systems and some architecture and visualization problems. So it was rewritten from scratch [39]. New version of library uses a similar application programming interface (API) for configuration of CGT and already existing CGT can be easily adopted to the new API. It actively uses WebGL and shaders, that allow to get the entire control of a graphic pipeline. Also it has modular and extensible architecture which is presented in Fig. 8.

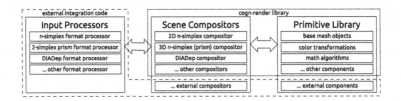

Fig. 8. Library architecture for cognitive tools visualization

The library consists of three component groups: Input Processors, which make preprocessing data from an intelligent system outputs and serves as integration layer into an intelligent system; Scene Compositors, which are responsible for creating, configuring and placing of primitives; and Universal Primitive Library, which contains a lot of common components used in creating any scene. As this version was developed based on web-technologies, it is very easy to integrate it to almost all intelligent systems, especially web-based ones.

Current version has the following advantages over the legacy one:

1. Ability to work on different platforms, easy integration into other systems.
2. Complete antialiasing for every visualized object that allows to render more attractive visualizations.
3. Fully functional Z-buffer that allows to fix a lot of fixed rendering issues which were in the legacy version.
4. Rendering of semi-transparent objects, that allows to implement a lot of new features.
5. Raycasting that allows to implement a lot of interactivity rich features.
6. Modular architecture and easy interface for extension.

For library integration into another system, it is necessary to write a small layer of an integration code preparing data from external system, and creating scene, and configuring appearance of the cognitive graphic tool. E.g. integration with Moodle requires implementation of the following items: (1) plugin for Moodle on PHP which extracts data from Moodle database and transfers them to a page; (2) input processor module on javascript which places html-element for the cognitive graphic tool, creating scene and transferring data to it.

It should be noted, that library is under rapid development and created components may be insufficient for all other intelligent systems demands. Because this library interface was made extendable, another developer can easily implement a component, requiring external code, and attach it to a standard library render pipeline. In future development of the library, the most popular and common modules will be integrated into the library repository in order to simplify integration and labor optimization for the future integrations into the new intelligent systems.

7   Concluding Remarks

The following new components for developed intelligent learning technology are proposed: an original approach to assess students' learning results, based on MDT; a new approach to the prediction of students' learning results, based on MDT and 2-simplex prism.

The original approach allows to assess student skills by independent learning axes revealing optimally students' level of different skills. The axes are the following: theory knowledge level, ability to practical tasks solving, ability of electric circuit design and research.

We propose the new approach to the prediction of students' learning results based on MDT and 2-simplex prism. Simple prototype of the prediction model for a learning process and cognitive graphic tools 2-simplex and 2-simplex prism, specificity of confidence region visualization are described. The proposed tools were verified on third year students' test results of speciality "Electrical Power Engineering". Variation of prediction polynome degree shows that for a majority of students prediction results obtained via using linear polynome are better than prediction results obtained via using square polynome. On this basis the following hypothesis can be formulated: the second derivative is not constant for a learning process and cannot be used for correct prediction. Confirmation of this hypothesis requires a full-scale research with more tests. Also a huge experiment base can be used for research of other polynomes or even other mathematical prediction models. In most cases, usage of the chosen simple prediction model shows appropriate quality of prediction, thus it will be used in the first version of an intelligent learning technology development. The proposed approach allows to construct individual learning trajectory for every student, predicting the priority skills which should be primarily developed. Interactive demonstration of cognitive graphic tool is prepared and published online.

The technology based on learning and testing intelligent systems are aimed at the future development, creating MDT tests for the course "Selected Chapters of Electronics", development and integration special Moodle plugins needed for the current research, collection huge students' learning results base, development and extension of software library for cognitive tools visualization.

References

1. Brusilovsky, P., Knapp, J., Gamper, J.: Supporting teachers as content authors in intelligent educational systems. Int. J. Knowl. Learn. **2**(3/4), 191–215 (2006). https://doi.org/10.1504/IJKL.2006.010992
2. Uskov, V., Uskov, A.: Innovative computer game technology curriculum. In: Proceedings of the 13th IASTED International Conference on Computers and Advanced Technology in Education, Lahaina, Maui, USA (2010). https://doi.org/10.2316/P.2010.709-067
3. Yankovskaya, A.: Logical tests and means of cognitive graphics. LAP LAMBERT Academic Publishing (2011). (in Russian)
4. Yankovskaya, A., Yevtushenko, N.: Finite state machine (FSM)-based knowledge representation in a computer tutoring system. In: Kommers, P., et al. (eds.) New Media and Telematic Technologies for Education in Eastern European Countries, pp. 67–74. Twenty University Press, Enshede (1997)
5. Singer, F.M., Stoicescu, D.: Using blended learning as a tool to strengthen teaching competences. Procedia Comput. Sci. **3**, 1527–1531 (2011). https://doi.org/10.1016/j.procs.2011.01.043
6. Bonk, C.J., Graham, Ch.R.: Handbook of Blended Learning: Global Perspectives, Local Designs, 624 p. Wiley, Hoboken (2006)
7. Stan, M.M.: The relationship of learning styles, learning behaviour and learning outcomes at the Romanian students. Procedia - Soc. Behav. Sci. **180**, 1667–1672 (2015). https://doi.org/10.1016/j.sbspro.2015.05.062

8. Shaidullin, R.N., Safiullin, L.N., Gafurov, I.R., Safiullin, N.Z.: Blended learning: leading modern educational technologies. Procedia - Soc. Behav. Sci. **131**, 105–110 (2014). https://doi.org/10.1016/j.sbspro.2014.04.087

9. Trends, E-Learning Market: Forecast 2014–2016 Report. A report by Docebo. http://www.docebo.com/landing/contactform/elearning-market-trends-and-forecast-2014-2016-docebo-report.pdf

10. Hattie, J., Yates, G.C.R.: Visible Learning and the Science of How We Learn, 368 p. Routledge, Abingdon (2013). https://doi.org/10.4324/9781315885025

11. Barana, A., Marchisio, M.: Ten good reasons to adopt an automated formative assessment model for learning and teaching mathematics and scientific disciplines. Procedia - Soc. Behav. Sci. **228**, 608–613 (2016). https://doi.org/10.1016/j.sbspro.2016.07.093

12. Samigulina, G., Samigulina, Z.: Intelligent system of distance education of engineers, based on modern innovative technologies. Procedia - Soc. Behav. Sci. **228**, 229–236 (2016). https://doi.org/10.1016/j.sbspro.2016.07.034

13. Harandi, S.R.: Effects of e-learning on students' motivation. Procedia - Soc. Behav. Sci. **181**, 423–430 (2015). https://doi.org/10.1016/j.sbspro.2015.04.905

14. Dzandu, M.D., Tang, Y.: Beneath a learning management system-understanding the human information interaction in information systems. Procedia Manuf. **3**, 1946–1952 (2015). https://doi.org/10.1016/j.promfg.2015.07.239

15. Andreicheva, L., Latypov, R.: Design of E-learning system: M-learning component. Procedia - Soc. Behav. Sci. **191**, 628–633 (2015). https://doi.org/10.1016/j.sbspro.2015.04.580

16. Urh, M., Vukovic, G., Jereb, E., et al.: The model for introduction of gamification into e-learning in higher education. Procedia - Soc. Behav. Sci. **197**, 388–397 (2015). https://doi.org/10.1016/j.sbspro.2015.07.154

17. Yankovskaya, A.E.: Design of optimal mixed diagnostic test with reference to the problems of evolutionary computation. In: Proceedings of the First International Conference on Evolutionary Computation and Its Applications, Moscow, pp. 292–297 (1996)

18. Yankovskaya, A.E., Semenov, M.E.: Intelligent system for knowledge estimation on the base of mixed diagnostic tests and elements of fuzzy logic. In: Proceedings of IASTED International Conference on Technology for Education (TE 2011), Dallas, USA, pp. 108–113 (2011). https://doi.org/10.2316/P.2011.754-001

19. Yankovskaya, A.E., Semenov, M.E.: Decision making in intelligent training-testing systems based on mixed diagnostic texts. Sci. Tech. Inf. Process. **40**(6), 329–336 (2013). https://doi.org/10.3103/s0147688213060087

20. Yankovskaya, A.E., Semenov, M.E.: Foundation of the construction of mixed diagnostic tests in systems for quality control of education. In: Proceedings of 13th IASTED International Conference Computers and Advanced Technology in Education (CATE 2010), Maui, Hawaii, USA, pp. 142–145 (2010)

21. Yankovskaya, A.E: Mixed diagnostic tests are a new paradigm of construction of intelligent learning and training systems. In: Proceedings of the New Quality of Education in the New Conditions, Tomsk, Russia, vol. 1, pp. 195–203 (2011). (in Russian)

22. Yankovskaya, A.E., Semenov, M.E.: Application mixed diagnostic tests in blended education and training. In: Proceedings of the IASTED International Conference Web-based Education (WBE 2013), Innsbruck, Austria, pp. 935–939 (2013). https://doi.org/10.2316/P.2013.792-037

23. Yankovskaya, A.E., Fuks, I.L., Dementyev, Y.N.: Mixed diagnostic tests in construction technology of the training and testing systems. Int. J. Eng. Innov. Technol. **3**(5), 169–174 (2013)
24. Yankovskaya, A., Dementyev, Y., Lyapunov, D., Yamshanov, A.: Intelligent information technology in education. In: Information Technologies in Science, Management, Social Sphere and Medicine (2016). https://doi.org/10.2991/itsmssm-16.2016.11
25. Yankovskaya, A., Razin, V.: Learning management system based on mixed diagnostic tests and semantic web technology. Tomsk State Univ. J. **2**(35), 78–98 (2016). https://doi.org/10.17223/19988605/35/9. (in Russian)
26. Yankovskaya, A., Dementyev, Y., Lyapunov, D., Yamshanov A.: Design of individual learning trajectory based on mixed diagnostic tests and cognitive graphic tools. In: Proceedings of the 35th IASTED International Conference Modelling, Identification and Control (MIC 2016), Innsbruck, Austria, pp. 59–65 (2016)
27. Yankovskaya, A., Yamshanov, A., Krivdyuk, N.: Application of cognitive graphics tools in intelligent systems. IJEIT **3**(7), 58–65 (2014)
28. Yankovskaya, A., Yamshanov, A.: Development of cross-platform cognitive tools invariant to problem areas and their integration into intelligent systems. Key Eng. Mater. **683**, 609–616 (2016). https://doi.org/10.4028/www.scientific.net/KEM.683.609
29. Yankovskaya, A.E., Yamshanov, A.V.: Application of 2-simplex prism for researching and modelling of processes in different problem areas. In: Proceedings of the Seventh International Conference on Cognitive Science, Svetlogorsk, Russia, pp. 655–657 (2016). (in Russian)
30. Yankovskaya, A., Yamshanov, A.: Family of 2-simplex cognitive tools and their applications for decision-making and its justification. In: Computer Science & Information Technology (CS & IT), pp. 63–76 (2016). https://doi.org/10.5121/csit.2016.60107
31. Yankovskaya, A., Levin, I., Fuks, I.: Assessment of teaching and learning by mixed diagnostic testing. In: Proceedings of the Frontiers in Mathematics and Science Education Research Conference (FISER-14), Famagusta, North Cyprus, pp. 86–93 (2014)
32. Yankovskaya, A.E., Semenov, M.E.: Construction of the mixed tests of the quality system of education. In: Proceedings of the International Scientific Conference (Modern IT& (e-) Training), Astrakhan, Russia, pp. 125–129 (2009). (in Russian)
33. Yankovskaya, A.E., Semenov, M.E., Yamshanov, A.V., Semenov, D.E.: Cognitive tools in learning and testing systems based on mixed diagnostic tests. Artif. Intell. Decis. Making **4**, 51–61 (2015). (in Russian)
34. Zhuravlev, Yu.I., Gurevitch, I.B.: Pattern recognition and image analysis. In: Pospelov, D.A. (ed.) Artificial Intelligence in 3 Books. Book 2: Models and Methods: Reference Book, pp. 149–191. Radio and Comm., Moscow (1990). (in Russian)
35. Yankovskaya, A.E.: Transformation of features space in patterns space on the base of the logical-combinatorial methods and properties of some geometric figures. In: Proceedings of the International Conference Pattern Recognition and Image Analysis: New Information, Abstracts of the I All-Union Conference, Part II, Minsk, pp. 178–181 (1991). (in Russian)
36. Kondratenko, S.V., Yankovskaya, A.E.: System of visualization TRIANG for decision-making justification with cognitive graphics usage. In: Proceedings of the Third Conference on Artificial Intelligence, vol. I, Tver, pp. 152–155 (1992). (in Russian)

37. Demo for Developed Cognitive Tool. http://cogntool.tsuab.ru/demos/2-simplex-prediction/

38. Source Code of Cognitive Tools Prototype Visualization Library. https://github.com/zZLOiz/cogn-proto

39. Source Code of Cognitive Tools Visualization Library. https://github.com/zZLOiz/cogn-render

Morphological and Technological
Approaches to Image Analysis

Application of Information Redundancy Measure To Image Segmentation

Dmitry Murashov$^{(\boxtimes)}$

Federal Research Center "Computer Science and Control" of RAS, Moscow, Russia
d_murashov@mail.ru

Abstract. In this paper, the problem of image segmentation quality is considered. The main idea is to find a quality criterion, which could have an extremum. The problem is viewed as selecting the best segmentation from a set of images generated by segmentation algorithm at different parameter values. We propose to use information redundancy measure as a criterion for optimizing segmentation quality. The method for constructing the redundancy measure provides criterion with extremal properties. To show efficiency of the proposed criterion, computing experiment is carried out. The proposed criterion is combined with SLIC and EDISON segmentation algorithms. Computing experiment shows that the segmented image corresponding to a minimum of redundancy measure produces acceptable information distance when compared with the original image. In most cases, the lowest information distance between this segmented image and ground-truth segmentations is obtained. An example of applying the redundancy measure to segmentation of images of painting material cross-sections is considered.

Keywords: Image segmentation · Segmentation quality · Redundancy measure · Variation of information · Painting material cross-sections

1 Segmentation Quality Problem

The paper deals with the problem of image segmentation quality. According to Haralick and Shapiro [16], segmentation is the process of partitioning an image into non-overlapping subregions. The elements in subregions are grouped by some feature and differ from the elements of the adjacent areas. Formal definition of segmentation is given in [14]. Any of the segmentation algorithms has one or more parameters. Parameters should be set in order to provide the best quality of the segmentation result. The problem of setting parameters is rather difficult. In this work we formulate the problem of segmentation quality as follows. For a given input image U varying parameter t of the segmentation algorithm, we obtain a set of J segmented images $\mathcal{V} = \{V_1, V_2, \ldots, V_j, \ldots, V_J\}$. It is necessary to choose an image V_j giving minimum to performance criterion $M(U, V_j)$:

$$V_{\min} = \operatorname*{arg\,min}_{V_j} \left(M\left(U, V_j\right) \right), j = 1, 2, \ldots, J. \tag{1}$$

© Springer Nature Switzerland AG 2019
V. V. Strijov et al. (Eds.): IDP 2016, CCIS 794, pp. 125–139, 2019.
https://doi.org/10.1007/978-3-030-35400-8_9

When solving different tasks of image analysis, an estimation of quality should be made. This may be a visual evaluation of an expert or computation of any quantitative measure. The segmentation results are usually compared with an image partitioned manually and accepted as ground-truth [4]. The quality can be represented by parameters describing boundary detection error, region consistency, and segment covering. In papers [20] and [4], the authors used precision-recall framework for comparing segment boundaries. In [19] Martin et al. proposed global and local consistency errors as the measures for comparing segments in images generated by segmentation algorithm and in ground-truth segmentations. Some other measures for evaluating segmentation quality are discussed in works [27] and [10].

If the segmentation operation is considered as a clustering of pixels, then the set-theoretical, statistical, and measures in terms of information theory are usually applied [26]. The most commonly used are: chi-square measure; Rand Index [24] and its variants [25]; Fowlkes-Mallows measure [12]; mutual information and normalized mutual information [3]; variation of information [23]. These measures allow comparing different versions of partitioning image into non-overlapping regions. In the paper [4], the authors noted that the standard methodology for estimating efficiency of segmentation algorithms has not been developed yet.

In the paper [13], another approach is proposed. Parameter of the superpixel segmentation algorithm [11] was chosen depending on the result of estimating similarity between segmented and original images. As a measure of similarity the authors proposed to use weighted uncertainty index computed using the values of the normalized mutual information [3] between the color channels of the original and segmented images. Frosio and Ratner proposed to choose parameter value providing the best segmentation in terms of visual perception. The dependence of uncertainty index on parameter value (and accordingly, the number of subregions) is approximately monotonous (see [13]). Using expert estimations of segmentations at various parameter values, the areas of under-segmentation, over-segmentation, and optimal segmentation were outlined in the space "parameter – uncertainty index". At the processing phase, an iterative procedure was applied to obtain parameter value providing the best image partition. This method drawbacks are related to the necessity of training procedure. Segmentation algorithm produces acceptable results only for those types of images which were involved in the training process.

If the segmentation method is developed according to the criteria of visual perception, then a human visual system model should be used. It is also preferable to have peaked or concave dependency of the segmentation system's performance index on the parameter value.

In the work [5], a theoretic-information model of the human visual system is proposed. The model is based on Barlow's hypothesis [6] about minimizing data redundancy at the early stages of signal processing in the human visual system.

In this work, based on principle of data redundancy minimization [5], we propose to use measure of information redundancy as a segmentation quality criterion. We show that particular method of forming theoretic-information criterion provides it with extremum. This criterion can be used for developing unsupervised segmentation techniques. We demonstrate that the proposed criterion can be combined with various segmentation algorithms for developing unsupervised segmentation techniques. Then a fusion of SLIC algorithm and the redundancy measure is applied to segmenting layers in images of painting material cross-sections.

2 Optimization of Segmentation Quality

To study the problem (1), segmentation process can be described by a model:

$$\mathbf{V} = \mathbf{F}(\mathbf{U}, \mathbf{t}), \tag{2}$$

where $\mathbf{U}$ is an input image represented as a vector, $\mathbf{V}$ is a segmented image, $\mathbf{F}$ is a vector-function (non-linear in general case) describing segmentation algorithm, $\mathbf{t}$ is a vector of parameters. Varying coordinates of parameter vector, one will obtain a set of segmented images. The result of segmentation depends on the applied algorithm parameter. It was stated above that the performance index of the system (2) will be taken in the form of theoretic-information measure. To apply theoretic-information approach, a probabilistic model of relationship between input and segmented images is needed. Segmentation quality will be estimated using one of the color channels (for example, L) of images in the CIE Lab color space.

Let the initial and segmented images be the input and the output of a stochastic information system. Levels of lightness in images are continuous random scalar variables U and V with probability mass functions of $p(u)$ and $q(v)$, where u and v are the values of U and V, respectively. Segmentation operation can be represented by an information channel model:

$$V = F(U + \eta, t), \tag{3}$$

where U is an input signal, V is a channel output, F is a transformation function, t is a parameter, and η is a channel noise. We assume that noise η is Gaussian random variable with zero mean value and variance σ_η^2; variables V and η are independent.

We propose to use a redundancy measure as a criterion of segmentation quality. The redundancy measure is defined as follows [5]:

$$R = 1 - \frac{I(U;V)}{C(V)}, \tag{4}$$

where $I(U;V)$ is mutual information between the system input and output, $C(V)$ is a channel capacity. We take $C(V) = H(V)$, where $H(V)$ is an entropy of the output. Taking into account the fact $I(U;V) = H(V) - H(V|U)$, we obtain

$$R = \frac{H(V|U)}{H(V)}, \tag{5}$$

where $H(V|U)$ is a conditional entropy of the output V under condition that the input is equal to U.

We show that the redundancy measure of the segmentation system described by the model (3–5) depends on the number of segments and can have a minimum. Probability mass function of the output may be represented by a sum

$$p(v) = \sum_{k=1}^{K} P(v_k)\,\delta(v - v_k), \tag{6}$$

where $P(v_k)$ is the probability of lightness value v_k associated with segment k, $\delta(v - v_k)$ is a delta-function, K – is the number of segments in the output image.

To find analytic dependence, we use a continuous version of the model (3). Taking into account the expression (6), differential entropy of the output can be written as follows:

$$H(V) = -\int_{-\infty}^{+\infty} p(v)\log p(v)\,dv = -\sum_{k=1}^{K} P(v_k)\log P(v_k). \tag{7}$$

Let all values of V be equiprobable: $P(v_k) = 1/K$. Then it follows from (7) that:

$$H(V) = \log K. \tag{8}$$

Next, we must find an expression for differential conditional entropy $H(V|U)$. Conditional entropy $H(V|U)$ is the measure of information about signal noise η measured at the system output. In this case, we may take [17]:

$$H(V|U) = H(\eta). \tag{9}$$

Differential entropy of the Gaussian noise is equal to:

$$H(\eta) = \frac{1}{2}\left[\log e + \log\left(2\sigma_\eta^2\right)\right], \tag{10}$$

where σ_η^2 is a variance of the system noise [17].

We assume that the probability mass function of the input image lightness is represented as a Gaussian mixture model of K components, which may overlap partially. The components of the mixture correspond to the segments of the output image V. Areas of component overlappings generate noise η. The overlapping areas are formed by pixels of U having the same lightness values, but related to different segments in image V. Substituting (8–10) into (5), we get the following expression for the redundancy measure:

$$R = \frac{\log e + \log\left(2\sigma_\eta^2\right)}{2\log K}. \tag{11}$$

It follows from (11) that the redundancy measure depends linearly on logarithm of system noise variance and inversely on logarithm of outlined segments number K. Function (11) has minimum at a point $K_{\min}$ if the noise variance σ_η^2 will be close to zero at small K and will rapidly grow when K increases. Computing experiments confirmed that such behavior of noise variance takes place.

Based on the proposed criterion, an unsupervised segmentation technique can be constructed in the following way. Suppose, chosen segmentation algorithm has parameter t affecting the partitioning (see model (3)). The input image U is segmented at different parameter values. As a result, a set of J segmented images $\mathcal{V} = \{V_1, V_2, \cdots, V_J\}$ is obtained. Next, for input image U and each of the segmented images V_j, the redundancy measure R is computed. Image V_j, which provides minimum to R: $R(U, V_j) = R_{\min}$, should be chosen. This image partitioned into $K_{\min}$ segments, fits parameter value $t = t_{\min}$.

In the next section, to illustrate the efficiency of applying the proposed criterion to image segmentation, computing experiment is performed.

3   Computing Experiment

The experiment includes four stages, and the main goals are as follows. Firstly, it will be confirmed that the redundancy measure $R(U, V_j)$, $V_j \in \mathcal{V}$, has minimum. Secondly, it will be shown that the dissimilarity between segmented image V_{min} and input image U is relatively small. Thirdly, it will be shown that the dissimilarity between segmented image V_{min} and ground-truth segmentations of input image U is less than the dissimilarity between any other $V_j \in \mathcal{V}$ and ground-truth segmentations. At last, we will demonstrate that the proposed criterion can be combined with alternative segmentation techniques.

In this section, the specific information redundancy measure (5) proposed in the previous section for choosing the best segmentation is used with superpixel algorithm SLIC (Simple Linear Iterative Clustering) [1,2] equipped with a post-processing procedure. SLIC algorithm partitions image into multiple small segments, or so-called superpixels. Post-processing procedure is applied for merging neighboring superpixels to produce homogeneous regions corresponding to the objects visible in the original image. For making a decision on merging, a threshold decision rule is used. This rule will allow merging if the following inequality takes place:

$$d(C_iC_j) \leq T(t), \tag{12}$$

where $d(C_iC_j)$ is the distance between centers of adjacent superpixels with numbers i and j in the selected color space; T is a threshold value, t is a parameter of model (3); $T = g(t)$. We take threshold value in the form of $T = g(t)$; $g(t)$ is a function, nonlinear in general case. Here, we use $T = g_c t$, $g_c = \text{const} = 1$.

In the experiments, we used 33 images from the Berkeley Segmentation Dataset BSDS500 [4] transformed to CIE Lab color space. Two of these test images are shown in Fig. 1.

At the first stage, each one of test images is segmented using algorithm SLIC and post-processing procedure at different values of parameter t. It is necessary

(a)   (b)

Fig. 1. Test images taken from BSDS500 dataset

to show that the redundancy measure has a minimum. Each image generates a set of J segmented images $\mathcal{V} = \{V_1, V_2, \cdots, V_J\}$. For input image U and each of the segmented images V_j, the redundancy measure R is computed. To involve all color channels, we use the weighted version of the redundancy R_W:

$$R_{\mathrm{W}}(U, V_j) = \frac{R_{\mathrm{L}}H_{\mathrm{L}}(U) + R_{\mathrm{a}}H_{\mathrm{a}}(U) + R_{\mathrm{b}}H_{\mathrm{b}}(U)}{H_{\mathrm{L}}(U) + H_{\mathrm{a}}(U) + H_{\mathrm{b}}(U)}, \tag{13}$$

where R_i is the redundancy measure determined in color channel $i \in \{L, a, b\}$ of images U and V_j; H_i is the entropy of the color channel i of the input image.

We apply SLIC algorithm and the post-processing procedure to all test images. For each test image a set of segmented images is generated at initial superpixel size of $a = 16$ pixels, parameter $m = 2$ (for details see [1], [2]), and parameter of the postprocessing procedure t is changing in the range $0 \leq t \leq 3.6$ with increment equal to 0.2. Relationship between parameter t and the number of segments K in images V_j generated by one of the test images is shown in Fig. 2. The curve demonstrates one-to-one correspondence between t and K. Therefore, we consider the function of number of segments K instead of t.

At the second stage, segmentation quality is estimated. We estimate the amount of input image information which was lost in segmentation process. For this purpose we compare segmented images with the input image U using normalized version of variation of information proposed in [22,23] for comparing

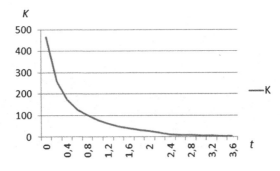

Fig. 2. Relationship between parameter t value and number of segments K

clusterings. We use this metric because normalized variation of information does not depend on the pixel number in an image. This metric was also used in [4] for comparing segmented images. Here we use the weighted index based on this metric:

$$VI_W(U, V_j) = \frac{VI_L H_L(U) + VI_a H_a(U) + VI_b H_b(U)}{H_L(U) + H_a(U) + H_b(U)}, \tag{14}$$

$$VI_i(U, V_j) = \frac{H_i(U) + H_i(V_j) - 2I_i(U; V_j)}{H_i(U, V_j)}, \tag{15}$$

where VI_W is the weighted variation of information between U and V_j; VI_i is the distance between color channels i of images U and V_j; $I_i(U; V_j)$ is the mutual information; $H_i(U, V_j)$ is the joint entropy.

At the third stage, using the weighted index (14) based on metric (15), we compare a set of J segmented images with the ground-truth segmentations V_s^{GT}, $s = 1, 2, \ldots, S$, where S is a number of ground-truth segmentations of a test image U available in BSDS500 dataset.

To show that application domain of the proposed quality criterion is not restricted only to a particular technique, at the last stage of the experiment we combine redundancy criterion with mean shift based EDISON segmentation algorithm presented in [8,9,21]. EDISON algorithm is also applied to images from BSDS500 dataset.

4   Results

To demonstrate results of the experiment, the images shown in Fig. 1 are used. At the experiment first stage, a set of segmented images is generated from each of the test images. For a test image and related set of segmented images we have computed the weighted redundancy measure R_W. Dependency of measure R_W on the number of segments K for the images shown in Fig. 1 is depicted in Fig. 3. Minimum of R_W is reached at values of $K = 87$ and $K = 25$ corresponding to parameter values $t = 2$ and $t = 2.4$ respectively.

Next, in order to estimate the distance between the input and the segmented images, we compute weighted normalized variation of information (14, 15). The curves representing VI_W as the functions of number of segments K are depicted in Fig. 3(a,b) by dashed lines. One can see that distance between the input and segmented images decreases when K grows and become nearly stable at K_{min} corresponding to minimal redundancy value.

At the next stage we compare generated sets of segmented images with ground-truth segmentations. The results of comparison obtained for the images shown in Fig. 1 are represented in Fig. 4 as the curves reflecting relationship between normalized variation of information VI_W and number of segments K in images V_j. It can be seen from Fig. 4 that for the majority of the ground-truth segmentations, the distance is minimal when an image is partitioned into 55 segments. At this K value the redundancy measure reaches minimum. Taking

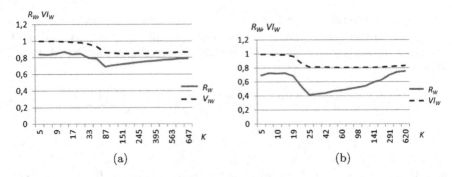

(a)   (b)

Fig. 3. Dependency of redundancy R_W and normalized variation of information VI_W on the number of segments K for images shown in Fig. 1

into account the fact that ground-truth segmentations were produced manually, we can conclude that the proposed technique allows one to obtain the best segmentation in terms of visual perception.

Ground-truth segmentations of images shown in Fig. 1(a), (b) are shown in Fig. 5(a) and (c) respectively. Segmented images fitting the minimum condition of the redundancy measure are depicted in Fig. 5(b) and (d). It can be seen from Fig. 5 that key details of the original images are captured in the segmented images as well as in the ground-truth segmentations.

To show efficiency of the proposed technique, we introduce the following relative difference:

$$\Delta K_{\mathrm{rel}} = \frac{K_{\min} - K_{\min}^{\mathrm{GT}}}{K_{\max}}, \tag{16}$$

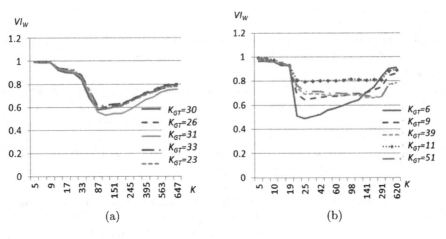

(a)   (b)

Fig. 4. Normalized variation of information $VI_W\left(V_j, V_s^{\mathrm{GT}}\right); j = 1, 2, \ldots, J; s = 1, 2, \ldots, S$, computed for segmented images $V_j \in \mathcal{V}$ and ground-truth segmentations with different number of segments K_{GT}

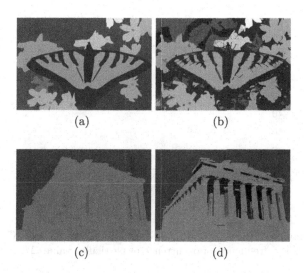

(a)   (b)

(c)   (d)

Fig. 5. Segmented and ground-truth images: (a) ground-truth segmentation of image shown in Fig. 1(a), $K_{GT} = 33$; (b) segmented image, $K = 87$; (c) ground-truth segmentation of image shown in Fig. 1(b), $K_{GT} = 6$; (d) segmented image, $K = 25$

where $K_{\min}$ is a number of segments corresponding to $R_{\min}$; $K_{\min}^{GT}$ is a number of segments in image V_j, which corresponds to the minimum of distance $VI_W(V_s^{GT}, V_j)$; $K_{\max}$ is the greatest possible number of segments in images $V_j, j = 1, 2, \ldots, J$ obtained from input image U. For example, for image shown in Fig. 1(b), we have the following values of variables in (16): $K_{\min} = 25$, $K_{\max} = 620$, and $K_{\min}^{GT} = 25$ for the ground-truth segmentations with number of segments $K_{GT} = 6$, $K_{GT} = 9$, and $K_{GT} = 11$; $K_{\min}^{GT} = 183$ for the ground-truth segmentations with $K_{GT} = 39$ and $K_{GT} = 11$ (see Fig. 5(c)). Histogram of ΔK_{rel} values computed for 33 test images and 165 ground-truth segmentations (5 ground-truth segmentations per each test image) is depicted in Fig. 6. Figure 6 shows that there exists a sufficiently large group of images which produce rather small magnitude of ΔK_{rel}. The ground-truth segmentations of these images are close enough in the sense of measure (14–15) to the segmentations minimizing the redundancy of information.

At the last stage of the experiment, weighted version (13) of the proposed criterion (5) is applied to EDISON segmentation algorithm. We set the following parameter values: feature (color) bandwidth is equal to 6.5, spatial bandwidth is equal to 7, and minimum region area (in pixels) S_{reg} runs from 200 to 10000. Here, we give results obtained using the image depicted in Fig. 1(a). For different values of S_{reg}, a set of segmentations V_j is generated and values of measures $R_W(U, V_j)$ and $VI_W(U, V_j)$ are computed and presented in Fig. 7(a, b). Minimum of the redundancy measure $R_{W\min} = 0.712$ is reached at $S_{reg} = 1300$ corresponding to number of segments $K_{\min} = 37$. The distance measured by weighted variation of information between the input and segmented images decreases from $VI_W(U, V_j) = 0.95$ to $VI_W(U, V_j) = 0.878$ when K grows

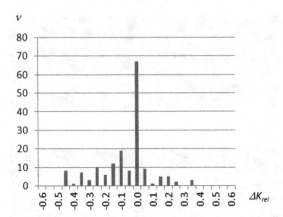

Fig. 6. Histogram of ΔK_{rel} values computed for 33 test images and 165 ground-truth segmentations; ν is a frequency of occurrence of particular value ΔK_{rel}

and become nearly stable at $K_{\min} = 37$ corresponding to minimal redundancy value (see Fig. 7(a)). For three of five ground-truth segmentations, the measure of dissimilarity $VI_W(V_s^{\mathrm{GT}}, V_j)$ reaches minimum when the input image is partitioned into 37 segments. At this K value the redundancy measure R_W reaches minimum (see Fig. 7(b)).

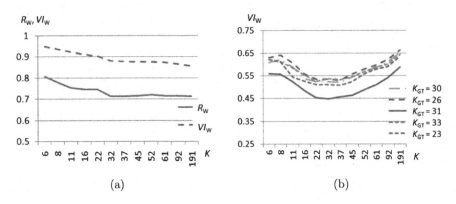

(a)   (b)

Fig. 7. Segmentation results obtained using EDISON algorithm: (a) redundancy measure R_W and normalized variation of information VI_W represented as the functions of number of segments K for image shown in Fig. 1 (a); (b) normalized variation of information $VI_W\left(V_j, V_s^{\mathrm{GT}}\right), j = 1, 2, \ldots, J$, and $s = 1, 2, \ldots, S$, computed for segmented images $V_j \in \mathcal{V}$ and ground-truth segmentations V_s^{GT} with different number of segments K_{GT}

The segmented image obtained using EDISON algorithm is shown in Fig. 8. This part of the computing experiment shows that the proposed criterion of segmentation quality can be successfully used for segmenting images together with various algorithms.

Fig. 8. The result of segmenting image shown in Fig. 1 (a) using EDISON algorithm, $K_{min} = 37$

5   Segmenting Images of Artwork Material Cross-sections

In this section we apply developed technique to the task of segmenting microscopic images of artwork's ground and paint cross-sections.

One of the main parts of the study aimed on the attribution of artworks created in the technique of easel oil painting is the stratigraphic and morphological analysis of paint layers. An important step for obtaining stratigraphic information is the precise paint and ground layer segmentation in microscopic images of cross-sections. The properties of the painting cross-sections produce difficulties for segmentation task.

For segmenting paint layers, several techniques were proposed in literature. One of the known methods is based on k-means clustering of the set of color channels of specimen images in visual and ultraviolet spectral bands plus spatial information (x and y coordinates are included as another two channels) [7,18]. The number of classes is set a priori as a maximum expected number of layers by the user. The authors note that without expert knowledge the segmentation remains preliminary. In [28] for layer segmentation the authors used method proposed by Haindl et al. [15]. The segmentation is based on a weighted combination of several unsupervised segmentation results, each in different resolution. Images obtained in visual and ultraviolet spectral bands are locally represented by random field models. Single-resolution segmentation is based on the Gaussian mixture model. The authors consider that the segmentation results can be possibly modified by the experienced operator [28].

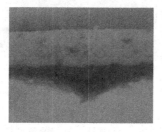

Fig. 9. Image of ground cross-section

In this work, the proposed in Sect. 2 technique based on the redundancy measure is applied to the task of segmenting paint and ground layers in microscopic images of painting material cross-sections. An image of ground cross-section is shown in Fig. 9. It can be seen that the cross-section put into specimen carrier has two ground layers and thin zone of interpenetration located between the layers. In this research, we use color JPEG images fixed by a digital camera Nikon Coolpix 4500 (4 MP) mounted on microscope MB10, and images taken by camera AxioCam ICc (3.3 MP) integrated in microscope Carl Zeiss Axio Imager 1. The images are acquired at magnification ×100. The size of the images is 2080 × 1540 pixels. In the research, the central image regions of size about 650 × 500 pixels and with rather small distortions are used. To reduce noise, the images are pre-processed by median filter. To extract paint layers, SLIC algorithm [1,2] with post-processing procedure based on threshold rule (12) is applied. Here, we use RGB color space. Color space choice is conditioned by the properties of cross-section images. The initial superpixel size a is taken equal to 12 pixels, increment of parameter t is equal to 0.05. A set of segmented images V_j is obtained at parameter t changing in the range $0 \leq t \leq 1.6$. For input image U shown in Fig. 9 and segmented images V_j, $0 \leq j \leq 32$, we computed R_W and VI_W using formulas (13–14). The curves R_W and VI_W represented as the dependencies on the number of segments K are shown in Fig. 10. Minimum of R_W corresponds

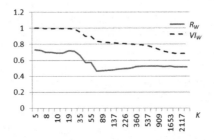

Fig. 10. Redundancy R_W and normalized variation of information VI_W curves obtained from image of ground cross-section shown in Fig. 9

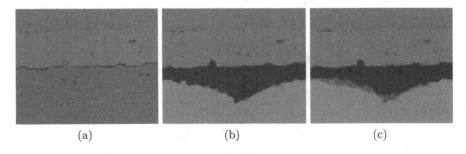

(a)   (b)   (c)

Fig. 11. Segmented images obtained at parameter t values equal to 0.85, 0.8, and 0.75 having 55, 64 and 89 segments, respectively

to the number of segments $K = 64$ and parameter value $t = 0.8$. In Fig. 11(a–c), segmented images of cross-section (see Fig. 9) obtained at $t = 0.85$, $t = 0.8$, and $t = 0.75$ containing 55, 64, and 89 segments respectively, are shown. Ground layers and specimen carrier material in Fig. 11(b) and (c) are represented as the largest segments.

6 Conclusions

The problem of image segmentation quality was considered. The problem was studied as a task of selecting the best segmentation from a set of images generated by segmentation algorithm at different parameter values. We proposed using information redundancy measure as a performance criterion. It was shown that the proposed way of constructing the redundancy measure provides the performance criterion with extremum. A technique based on theoretic-information criterion was proposed for selecting the best segmented image. Computing experiment was conducted using images from the Berkeley Segmentation Dataset. The experiment confirmed that the segmented image, corresponding to a minimum of redundancy measure produces acceptable information dissimilarity when compared with the original image and gives the minimal distance from the majority of ground-truth segmentations available in BSDS500 database. In the experiment, SLIC algorithm with the post-processing procedure and EDISON algorithm were involved for generating sets of partitioned images. The proposed segmentation quality criterion can be utilized with other segmentation algorithms for developing unsupervised segmentation techniques. As an example, the minimum redundancy criterion was successfully applied to the task of segmenting ground layers in microscopic images of painting material cross-sections.

Acknowledgements. The research was supported in part by the Russian Foundation for Basic Research (grants No. 18-07-01385 and No. 18-07-01231).

References

1. Achanta, R., Shaji, A., Smith, K., Lucchi, A., Fua, P., Süsstrunk, S.: Slic superpixels. Technical report (2010)
2. Achanta, R., Shaji, A., Smith, K., Lucchi, A., Fua, P., Süsstrunk, S.: Slic superpixels compared to state-of-the-art superpixel methods. IEEE Trans. Pattern Anal. Mach. Intell. **34**(11), 2274–2282 (2012). https://doi.org/10.1109/tpami.2012.120
3. Ana, L.N.F., Jain, A.K.: Robust data clustering. In: 2003 Proceedings of IEEE Computer Society Conference on Computer Vision and Pattern Recognition, vol. 2, pp. 128–133. IEEE (2003). https://doi.org/10.1109/cvpr.2003.1211462
4. Arbelaez, P., Maire, M., Fowlkes, C., Malik, J.: Contour detection and hierarchical image segmentation. IEEE Trans. Pattern Anal. Mach. Intell. **33**(5), 898–916 (2011). https://doi.org/10.1109/tpami.2010.161
5. Atick, J.J., Redlich, A.N.: Towards a theory of early visual processing. Neural Comput. **2**(3), 308–320 (1990). https://doi.org/10.1162/neco.1990.2.3.308

6. Barlow, H.B.: Possible principles underlying the transformations of sensory messages. Sens. Commun., 217–234 (1961). https://doi.org/10.7551/mitpress/9780262518420.003.0013

7. Beneš, M., Zitová, B., Hradilová, J., Hradil, D.: Image processing in material analyses of artworks. In: Proceedings of International Conference on Computer Vision Theory and Applications (VISAPP), pp. 521–524 (2008). https://doi.org/10.5220/0001079705210524

8. Christoudias, C.M., Georgescu, B., Meer, P.: Synergism in low level vision. In: 16th International Conference on Pattern Recognition, ICPR 2002, Quebec, Canada, 11–15 August 2002, pp. 150–155 (2002). https://doi.org/10.1109/ICPR.2002.1047421

9. Comaniciu, D., Meer, P.: Mean shift: a robust approach toward feature space analysis. IEEE Trans. Pattern Anal. Mach. Intell. 24(5), 603–619 (2002). https://doi.org/10.1109/34.1000236

10. Csurka, G., Larlus, D., Perronnin, F., Meylan, F.: What is a good evaluation measure for semantic segmentation? In: BMVC, vol. 27, pp. 32.1–32.11 (2013). https://doi.org/10.5244/c.27.32

11. Felzenszwalb, P.F., Huttenlocher, D.P.: Efficient graph-based image segmentation. Int. J. Comput. Vision 59(2), 167–181 (2004). https://doi.org/10.1023/B:VISI.0000022288.19776.77

12. Fowlkes, E.B., Mallows, C.L.: A method for comparing two hierarchical clusterings. J. Am. Stat. Assoc. 78(383), 553–569 (1983). https://doi.org/10.1080/01621459.1983.10478008

13. Frosio, I., Ratner, E.R.: Adaptive segmentation based on a learned quality metric. In: VISAPP 2015, vol. 1, pp. 283–292 (2015). https://doi.org/10.5220/0005257202830292

14. Gonzalez, R.C., Woods, R.E.: Digital Image Processing. Pearson Prentice Hall, Upper Saddle River (2008)

15. Haindl, M., Mikeš, S., Pudil, P.: Unsupervised hierarchical weighted multi-segmenter. In: Benediktsson, J.A., Kittler, J., Roli, F. (eds.) MCS 2009. LNCS, vol. 5519, pp. 272–282. Springer, Heidelberg (2009). https://doi.org/10.1007/978-3-642-02326-2_28

16. Haralick, R.M., Shapiro, L.G.: Image segmentation techniques. Comput. Vis. Graph. Image Process. 29(1), 100–132 (1985). https://doi.org/10.1016/s0734-189x(85)90153-7

17. Haykin, S.: Neural Networks: A Comprehensive Foundation, 2nd edn. Prentice Hall PTR, Upper Saddle River (1998)

18. Kaspar, R., Petru, L., Zitová, B., Flusser, J., Hradilova, J.: Microscopic cross-sections of old artworks. In: 2005 IEEE International Conference on Image Processing, ICIP 2005, vol. 2, pp. II-578. IEEE (2005). https://doi.org/10.1109/icip.2005.1530121

19. Martin, D., Fowlkes, C., Tal, D., Malik, J.: A database of human segmented natural images and its application to evaluating segmentation algorithms and measuring ecological statistics. In: 2001 Proceedings of Eighth IEEE International Conference on Computer Vision, ICCV 2001, vol. 2, pp. 416–423. IEEE (2001). https://doi.org/10.1109/iccv.2001.937655

20. Martin, D.R., Fowlkes, C.C., Malik, J.: Learning to detect natural image boundaries using local brightness, color, and texture cues. IEEE Trans. Pattern Anal. Mach. Intell. 26(5), 530–549 (2004). https://doi.org/10.1109/tpami.2004.1273918

21. Meer, P., Georgescu, B.: Edge detection with embedded confidence. IEEE Trans. Pattern Anal. Mach. Intell. **23**(12), 1351–1365 (2001). https://doi.org/10.1109/34. 977560
22. Meilă, M.: Comparing clusterings by the variation of information. In: Schölkopf, B., Warmuth, M.K. (eds.) COLT-Kernel 2003. LNCS (LNAI), vol. 2777, pp. 173–187. Springer, Heidelberg (2003). https://doi.org/10.1007/978-3-540-45167-9_14
23. Meilă, M.: Comparing clusterings: an axiomatic view. In: Proceedings of the 22nd International Conference on Machine Learning, pp. 577–584. ACM (2005). https:// doi.org/10.1145/1102351.1102424
24. Rand, W.M.: Objective criteria for the evaluation of clustering methods. J. Am. Stat. Assoc. **66**(336), 846–850 (1971). https://doi.org/10.2307/2284239
25. Unnikrishnan, R., Pantofaru, C., Hebert, M.: A measure for objective evaluation of image segmentation algorithms. In: Proceedings of the 2005 IEEE Computer Society Conference on Computer Vision and Pattern Recognition (CVPR 2005) - Workshops, CVPR 2005, vol. 03, pp. 34–41. IEEE Computer Society, Washington, DC (2005). https://doi.org/10.1109/CVPR.2005.390
26. Wagner, S., Wagner, D.: Comparing clusterings - an overview. Technical report 2006–04, Universität Karlsruhe (TH) (2007)
27. Zhang, H., Fritts, J.E., Goldman, S.A.: Image segmentation evaluation: a survey of unsupervised methods. Comput. Vis. Image Underst. **110**(2), 260–280 (2008). https://doi.org/10.1016/j.cviu.2007.08.003
28. Zitová, B., Beneš, M., Hradilová, J., Hradil, D.: Analysis of painting materials on multimodal microscopic level. In: IS&T/SPIE Electronic Imaging, pp. 75,310F–1– 75,310F–9. International Society for Optics and Photonics (2010). https://doi.org/ 10.1117/12.838872

Eyelid Position Detection Method for Mobile Iris Recognition

Gleb Odinokikh[1,3(✉)], Mikhail Korobkin[2], Vitaly Gnatyuk[3], and Vladimir Eremeev[3]

[1] Federal Research Center "Computer Science and Control" of the Russian Academy of Sciences, Moscow, Russia
g.odinokikh@gmail.com
[2] National Research University of Electronic Technology, Zelenograd, Russia
[3] Samsung R&D Institute Russia (SRR), Moscow, Russia

Abstract. Information about eyelid position in an image is used during iris recognition for eyelid and eyelash noise removal, iris image quality estimation and other purposes. Eyelid detection is usually performed after iris-sclera boundary localization which is a fairly complex operation itself. If the authentication is working on a hand-held device, this order is not always justified, mainly because of the device limited performance, user interaction difficulties and highly variable environmental conditions. In this case the eyelid position information could be used to determine whether the image should be passed for the further complex processing operations. This paper proposes a method of eyelid position detection for iris image quality estimation and further complete eyelid border localization and compares its performance with several similar existing methods on four open datasets.

Keywords: Eyelid detection · Mobile biometrics · Iris recognition

1 Introduction

Recent years show increasing usage of biometric technologies in different areas. For example, biometric authentication is considered as a candidate to replace conventional authentication schemes using keys, smart cards, PINs or passwords, in order to provide more easy and secure access control to PCs or mobile devices, premises, cars, Internet, financial transactions and many others. Iris recognition has some advantages over other biometric technologies [1–3], that make it one of the most preferable technologies for mobile biometric applications.

The iris recognition with a mobile device is complicated for many reasons. The first one is the unconstrained environmental conditions that degrade iris recognition performance due to noise on a human iris, caused by various reflections and occlusions by eyelids and eyelashes [4]. The second one is the user's interaction difficulties. The iris image quality degrades due to the user's behavior features like winking, hand shaking, squinting, gaze direction and so on. Some

© Springer Nature Switzerland AG 2019
V. V. Strijov et al. (Eds.): IDP 2016, CCIS 794, pp. 140–150, 2019.
https://doi.org/10.1007/978-3-030-35400-8_10

examples of the unconstrained images are shown in Fig. 1. And one more reason is that the application should provide a real-time performance on the device with limited computational and memory resources.

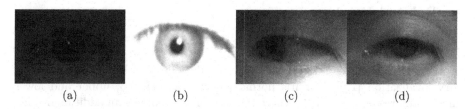

(a)   (b)   (c)   (d)

Fig. 1. Iris image quality degradation caused by human behavior and/or changing environmental condition such as poor illumination (a), over-exposure (b), gaze-away (c), eyelid occlusion (d)

This work focuses on the eyelid position detection during the iris recognition procedure. We define eyelid position as a distance to the eyelid from a pupil center by vertical (see Fig. 2). A definition of the points E_u and E_l is further described.

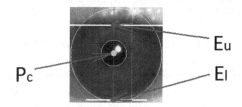

Fig. 2. Eyelid position determination, E_u, E_l are upper and lower eyelid position points respectively, P_c – pupil center

Based on the values of E_u and E_l, eye opening coefficient could be estimated that allows to decide, whether current frame should be skipped or passed to the next processing stages. If the frame passes this check, then it is possible to use the eyelid position information during the next iris-sclera, iris-eyelid border localization or eyelash detection stages. Otherwise, this information could be used to provide the user with feedback and to avoid the next operations proceeding to the next frame immediately.

The majority of the prior art methods aims to perform complete eyelid border localization. In order to compare them with the proposed method E_u and E_l values were calculated from the determined borders.

The proposed method is capable to perform accurate eyelid position detection right after the pupil detection step. It is robust and working fast enough to be applied on mobile device considering changes in environmental condition and user's behavior.

The paper is organized as follows: Sect. 2 describes the existing methods. In Sects. 3 and 4 the proposed method and experimental results are presented respectively. Conclusion is given in Sect. 5.

2   Related Work

2.1   Eyelid Detection Stage

In iris recognition eyelid detection stage usually follows either iris-sclera boundary localization [1,6–9] or iris normalization stage (Fig. 3, upper and lower image pairs respectively) [10], where the second one requires an additional time.

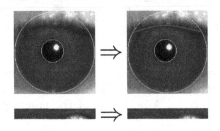

Fig. 3. Eyelid border localization stages

This paper is focused only on the methods running right after pupil or iris segmentation stages.

Conventional eyelid detection algorithm structure can be divided into two main parts: image preprocessing and eyelid localization. Often researchers propose just one of the steps and use the existing second one. Evaluation of the different proposed methods and preprocessing-localization combinations is performed. Methods showing best results are considered as final prior-art ones and are described below.

2.2   Eyelid Detection Methods

For eyelid border localization Daugman proposed to smooth the image with Gaussian filter and then use an Integro-differential operator (IDO) [1] for parabolic curve. Wildes proposed edge detection as the first step and parabolic Hough transform for eyelid localization [5]. Masek separated the eyelid and iris regions by horizontal lines after the edge detection step [11]. Kang and Park proposed to extract the eyelid candidate position with a local derivative mask and detect the eyelid region by curve fitting with IDO [12]. Zhang et al. proposed 1D peak shape filter to remove the eyelash noise and parabolic IDO on binary image for the localization [6]. Adam et al. in [7] and [13] applied anisotropic diffusion for denoising first, then extracted the eyelid edges and finally applied parabolic Hough transform for the localization. Yang et al. proposed asymmetric Canny operator for the eyelid edge extraction and parabolic curve fitting with least squares method for eyelid localization [14]. Kim, Cha et al. proposed eyelid detection method for gaze tracking purposes [15]. They proposed histogram equalization and further local minima searching for eyelid position detection. He et al. in [16] proposed 1D rank filter for eyelash noise removal, then horizontal edge detection and statistically established curvature models of upper and lower eyelids for the localization.

Almost all of the methods were proposed for and tested on the image datasets captured from the non-mobile device and therefore do not consider difficulties mentioned above. Most of them follow the iris-sclera boundary localization stage which is a fairly complex operation itself.

3   Eyelid Detection

In this paper a method of accurate eyelid position detection for the eye opening condition estimation is proposed. This estimation performs right after the pupil detection stage and could be used for making a decision about passing the frame for the further processing steps or providing user with feedback about his eye opening condition. The method is based on applying of multi-directional 2D Gabor filtering and 1D selective edge extraction as the preprocessing stage and sliding window approach for the eyelid detection as represented on the block diagram in Fig. 4.

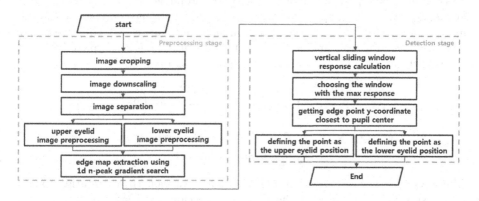

Fig. 4. Proposed algorithm block diagram

The choice of applying of two-dimensional Gabor filter with its shape and structure (see Fig. 6) was driven the following reasons: iris-eyelid border is a curve, that divide two areas of different intensity; it is also often characterized by the presence of eyelid shadow that become more pronounced with increasing of illumination. The filter is designed to emphasize the border using both edge and shadow information. Besides, chosen filter orientations allows to consider different eyelid shape and partially remove eyelash noise. Selective edge extraction is applied due to its capability to avoid the other noise factors like response from skin folds and horizontal eyelash noise. Thus, it makes the method more robust under unconstrained environmental conditions and user's behavior.

3.1 Preprocessing Stage

In accordance with the block diagram of algorithm (see Fig. 4) the preprocessing stage could be represented as in Fig. 6. A pre-cropped image is used as the input image for above diagram (see Fig. 6). Cropping parameters for the image are as follows: $width = n_W \cdot R_p$, $height = n_H \cdot R_p$, where R_p is the pupil radius, the image center corresponds to pupil center (X_p, Y_p) and n_W, n_H are two constants defining the image width and height respectively. In this work the values of $n_W = 4$ and $n_H = 12$, so $n_H/n_W = 3$ were picked up. These values were chosen experimentally and allow to consider eyelid position deviation in vertical direction and to restrict the region of search in horizontal direction. Despite the fact that pupil radius can change its value depending on many factors, the approach shows promising results (see Table 1) on all used testing datasets.

After the cropping, the image downscaling is performed with a scale factor equal to 0.5. This scale is sufficient to achieve the accuracy results represented in Table 1 while maintaining the required processing speed. The next step is separation for upper and lower eyelid image parts. There are two separation parameters defining the separated upper and lower eyelid image size: $U_{range}(0, Y_p + R_p)$ and $L_{range}(Y_p + R_p/2, Y_{max})$ as shown in Fig. 6, where zero corresponds to the first image row, and Y_{max} is equal to a number of image rows. After the upper and lower eyelid images are separated 2D Gabor filtering is performed for eyelid border extraction. The real part of the Gabor function is used:

$$g(x_g, y_g, \lambda, \theta, \psi, \sigma, \gamma) = \exp{-\frac{x_g'^2 + \gamma^2 y_g'^2}{2\sigma^2}} \cdot \cos\left(2\pi\frac{x_g'}{\lambda} + \psi\right),$$
$$x_g' = x_g \cos\theta + y_g \sin\theta,$$
$$y_g' = -x_g \sin\theta + y_g \cos\theta,$$

(1)

where $\lambda = 2\pi, \theta, \psi = \pi, \sigma = 3$ and $\gamma = 1$ are wavelength, orientation, phase offset, standard deviation of the Gaussian envelope and spatial aspect ratio of the envelope respectively. The same parameters of the filter, except orientation θ are used for all testing datasets. The kernel size is set to 15×15 pixels.

A convolution with Gabor filter is performed for both the upper and lower eyelid images for N_o different orientations (θ_n). In this work $N_o = 3$ and the orientations (clockwise) for the upper and lower eyelid images in degrees are:

$$\theta_{1...N}^{Upper} = 250.0, 270.0, 290.0,$$
$$\theta_{1...N}^{Lower} = 70.0, 90.0, 110.0.$$

(2)

These orientation values were obtained experimentally as giving the best results. Each eyelid image $I(x, y)$ is convolved with pre-defined kernel:

$$I'(x, y) = I(x, y) \otimes g'(x_g, y_g). \tag{3}$$

The kernel $g'(x_g, y_g)$ is combined from Gabor kernels of different θ_n as follows:

$$g'(x_g, y_g) = \frac{1}{N} \sum_{n=1}^{N} g(x_g, y_g, \theta_n). \tag{4}$$

The kernel examples for the different orientations θ_n are represented in Fig. 5.

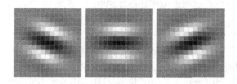

Fig. 5. Gabor kernel examples for different orientations (θ_n)

The convolution result image $I'(x, y)$ is also depicted in the scheme (see Fig. 6).

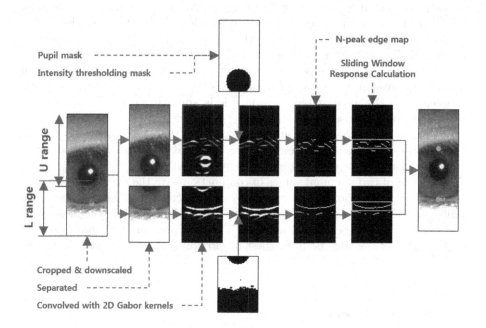

Fig. 6. Preprocessing stage scheme

As depicted in Fig. 6, the next step is overlaying of the mask $M(x, y)$ combined from two: pupil mask $M_P(x, y)$ and intensity mask $M_I(x, y)$ as follows

$$M(x, y) = M_p(x, y) \wedge M_I(x, y). \tag{5}$$

The pupil mask is determining area of pupil in the image. The intensity mask is determining overexposed image area with the next thresholding rule:

$$M_I(x, y) = \begin{cases} 255, & \text{if } I_s(x, y) < 240; \\ 0, & \text{otherwise,} \end{cases}$$

where $I_s(x, y)$ is a result of the image $I(x, y)$ convolution with Gaussian function kernel. The kernel parameters are used: size $= 3 \times 3$, $\sigma_0 = 1$. Combined mask $M(x, y)$ is dilated for a value $d = k_h/2 - 1$, where $k_h = k_w = 15$ is a dimension of the Gabor kernel.

The final step of the preprocessing is the edge map extraction. It is performed by searching of N_p largest local maxima values of gradient along column of each eyelid image $I'(x, y)$ outward from lower to upper boundary of upper eyelid image and from upper to lower boundary of lower eyelid image (see Fig. 6). In this work $N_p = 2$ and the gradient calculating window size is set to 3×1.

3.2 Detection Stage

A sliding window approach is used for eyelid position detection (see Fig. 6). The window response function reflects the dependence of number of the edge points inside the window from its vertical position. A dimension of the sliding window is set as depicted in Fig. 8, where 7 pixels by vertical was picked up as optimal since led to the best eyelid detection performance.

Further, two window locations (y_{max1}, y_{max2}) that correspond to maximal values of the response function $R(y)$ are considered as the candidates. To choose between them the next simple decision rule is applied (see Fig. 7), where T is a threshold value, chosen as $T = R(y_{max1})/3$ in this work.

The meaning of the values max_1 and max_2 is that max_1 is always farther from pupil center. And the scheme on Fig. 7 is valid for original image of lower eyelid and for upper eyelid image flipped by vertical.

Choosing of two maximal values in conjunction with the decision rule allows decreasing the detection errors caused by noise factors such as eyelashes, skin folds, glasses rims and others. After chosen the window location, the final eyelid position searching is performed inside it. The final position corresponds to vertical position of the edge point farthest from pupil center as depicted in Fig. 8.

4 Experimental Results

4.1 Accuracy Results

In order to show a performance of the proposed method, more than nine different existing methods have been implemented. The methods that have shown

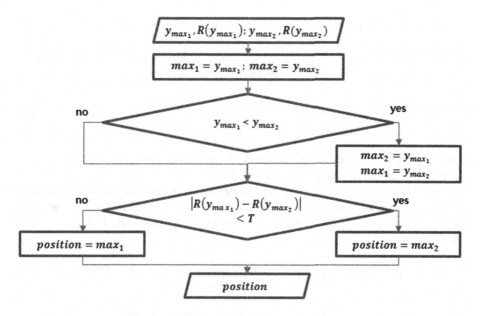

Fig. 7. Candidate position choosing rule

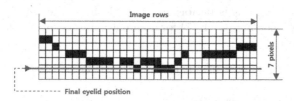

Fig. 8. Final eyelid position detection

the best results are considered as a prior art and used for comparison. All the methods including proposed are tested on four different datasets: CASIA-IrisV4-Thousand [17], CASIA-IrisV3-Lamp [18], AOPTIX [19] and MIR2016 (Train) dataset captured from mobile device [20]. All the datasets are labelled in the result tables (Table 1) as CS4, CS3, APX, and MIR respectively. More than 500 randomly selected images of each dataset were manually marked by expert for the evaluation. It is performed on the cropped images by marking E_u and E_l points y-coordinates (see Fig. 2).

The method of accuracy estimation with a set of certain admissible error values is chosen because it is justified in case of achieving better performance with a given precision is required. The precision of eyelid position detection is important for the next steps of iris recognition algorithm parts because it could be used to determination of searching regions for instance on further eyelid fitting or iris segmentation stages.

The detection accuracy evaluation is performed for certain value of admissible error ξ^{adm} equal to 5%. The relative error rate ε for the dataset is calculated as a power of the image set for those the error rate exceeds pre-defined threshold value corresponding to ξ^{adm} as follows:

$$\varepsilon = \frac{1}{N} \left| \{ \forall i : d_i > \xi^{adm} * height \} \right|,$$
$$d_i = |E(x, y_A)_i - E(x, y_M)_i|,$$

(6)

where $E(x, y_A)_i$ and $E(x, y_M)_i$ are eyelid points obtained after the algorithm applying and manually marked respectively for i-th image of one dataset, height – cropped image height (before downscaling) in pixels, N – total count of images in dataset. In order to show robustness of the method for different conditions and represent compactly results the accuracy value is averaged from all the testing datasets:

$$AVG_{\xi^{adm}} = \frac{100\%}{N_D} \sum_{i=1}^{N_D} (1.0 - \varepsilon_i),$$

(7)

where i is the dataset index, N_D is a count of testing datasets ($N_D = 4$ here).

In addition to the researchers methods evaluation for some preprocessing-localization combinations is performed. Table 1 shows the detailed information on the accuracy calculated as in (7) for different datasets considering $\xi^{adm} = 5\%$.

Table 1. Eyelid detection accuracy for $\xi^{adm} = 5\%$

Dataset	Upper eyelid					Lower eyelid				
	MIR	CS4	CS3	APX	AVG	MIR	CS4	CS3	APX	AVG
Daugman [1]	0.76	0.70	0.83	0.84	0.74	0.88	0.86	0.95	0.94	0.91
Wildes [5]	0.80	0.83	0.92	0.74	0.81	0.87	0.78	0.92	0.92	0.87
Masek [11]	0.50	0.70	0.90	0.93	0.73	0.40	0.65	0.86	0.95	0.72
Kang & Park [12]	0.86	0.89	0.90	0.88	0.86	0.96	0.88	0.95	0.94	0.93
Zhang et al. [6]	0.56	0.92	0.95	0.94	0.83	0.77	0.87	0.87	0.92	0.86
Adam et al. [7]	0.80	0.83	0.91	0.78	0.81	0.87	0.79	0.93	0.95	0.88
Yang et al. [14]	0.55	0.83	0.78	0.90	0.72	0.12	0.28	0.34	0.72	0.37
Kim et al. [15]	0.89	0.89	0.99	0.98	0.89	0.30	0.50	0.22	0.32	0.34
He et al. [16]	0.80	0.83	0.92	0.74	0.81	0.87	0.78	0.92	0.92	0.87
2DGF+IDO	0.93	0.90	0.95	0.91	0.92	0.97	0.86	0.92	0.96	0.93
Proposed	0.98	0.97	0.97	0.91	0.95	0.99	0.94	0.96	0.94	0.96

One more method showed good results is included: 2DGF+IDO is a combination of the proposed multi-directional 2D Gabor-filtering as the preprocessing and Daugman's integro-differential operator for the position detection.

4.2   Speed Measurement Results

The performance of the proposed method is measured on Samsung Galaxy Tab Pro 8.4 mobile device with Snapdragon 800 CPU (2.26 GHz Quad-core). The measurements are performed on the single core. Total execution time from getting of pupil data to both eyelids position detection is about 1.5–4 ms that suggests the possibility of using this method in real-time mobile applications. The measurements are not performed for other methods because it would require their additional optimizations and not guarantee achievement of implementation the results of researchers.

5   Concluding Remarks

The method of eyelid position detection is proposed. Contrary to existing approaches that require complete detection of eye features like iris-sclera border, iris-eyelid border this method can be used for fast input iris image quality estimation. This estimation can be used for the user feedback providing about his eye opening condition, further eyelid borders or eyelashes localization, iris segmentation parameters readjustment etc. The method showed promising results on accuracy (Table 1), and capability to work on a mobile device in real-time. The method robustness is demonstrated by its high accuracy results for all the tested datasets (Table 1). Aforementioned benefits make the method applicable for mobile iris recognition applications. Moreover, it should be noted that it shows good results on the low contrast images. Thus, possibly it could be used not only for the iris recognition, but in some other tasks required eye features processing like: gaze tracking, fatigue detection and so on.

References

1. Daugman, J.: How iris recognition works. IEEE Trans. Circuits Syst. Video Technol. **14**(1), 21–30 (2004). https://doi.org/10.1109/TCSVT.2003.818350
2. Chowhan, S., Shinde, G.: Iris biometrics recognition application in security management. In: Congress on Image and Signal Processing (CISP 2008), vol 1, pp. 661–665 (2008). https://doi.org/10.1109/CISP.2008.766
3. Bhattacharya, V., Mali, K.: Iris as a biometric feature: application, recognition, advantages & shortcomings. Int. J. Adv. Res. **3**(6), 1410–1415 (2013)
4. Dorairaj, V., Schmidt, N.A., Fahmy, G.: Performance evaluation of non-ideal iris based recognition system implementing global ICA encoding. In: IEEE International Conference on Image Processing (ICIP 2004), vol 3, pp. 11–14 (2004). https://doi.org/10.1109/ICIP.2005.1530384
5. Wildes, R.: Iris recognition: an emerging biometric technology. Proc. IEEE **85**(9), 1348–1363 (1997). https://doi.org/10.1109/5.628669
6. Zhang, X., Wang, Q., Zhu, H., Yao, C., Gao, L., Liu, X.: Noise detection of iris image based on texture analysis. In: Chinese Control and Decision Conference Proceedings (CCDC 2009), pp. 2366–2370 (2009). https://doi.org/10.1109/CCDC.2009.5192665

7. Adam, M., Rossant, F., Amiel, F., Mikovicova, B., Ea, T.: Reliable eyelid localization for iris recognition. In: Blanc-Talon, J., Bourennane, S., Philips, W., Popescu, D., Scheunders, P. (eds.) ACIVS 2008. LNCS, vol. 5259, pp. 1062–1070. Springer, Heidelberg (2008). https://doi.org/10.1007/978-3-540-88458-3_96
8. Gankin, K.A., Gneusev, A.N., Matveev, I.A.: Iris image segmentation based on approximate methods with subsequent refinements. J. Comput. Syst. Sci. Int. **53**(2), 224–238 (2014). https://doi.org/10.1134/S1064230714020099
9. Solomatin, I., Matveev, I.: Detecting visible areas of iris by qualifier of local textural features. J. Mach. Learn. Data Anal. **1**(14), 1919–1929 (2016). https://doi.org/10.21469/22233792.1.14.03
10. Min, T.-H., Park, R.-H.: Comparison of eyelid and eyelash detection algorithms for performance improvement of iris recognition. In: 15th IEEE International Conference on Image Processing (ICIP 2008), pp. 257–260 (2008) https://doi.org/10.1109/ICIP.2008.4711740
11. Masek, L.: Recognition of human iris patterns for biometric identification. Measurement **32**(8), 1502–1516 (2003)
12. Kang, B., Park, K.: A robust eyelash detection based on iris focus assessment. Pattern Recogn. Lett. **28**(13), 1630–1639 (2007). https://doi.org/10.1016/j.patrec.2007.04.004
13. Adam, M., Rossant, F., Amiel, F., Mikovikova, B., Ea, T.: Eyelid localization for iris identification. Radioengineering **17**(4), 82–85 (2008)
14. Yang, L., Wu, T., Dong, Y., Fei, L.: Eyelid localization using asymmetric Canny operator. In: Proceedings of International Conference on Computer Design and Applications, pp. 533–535 (2010)
15. Kim, H., Cha, J., Lee, W.: Eye detection for gaze tracker with near infrared illuminator. In: 17th IEEE International Conference on Computational Science and Engineering (CSE 2014), pp. 458–464 (2014). https://doi.org/10.1109/CSE.2014.111
16. He, Z., Tan, T., Sun, Z., Qiu, X.: Robust eyelid, eyelash and shadow localization for iris recognition. In: 15th IEEE International Conference on Image Processing (ICIP 2008), pp. 265–268 (2008). https://doi.org/10.1109/ICIP.2008.4711742
17. CASIA Iris Image Database V4.0. http://biometrics.idealtest.org/dbDetailForUser.do?id=4
18. CASIA Iris Image Database V430. http://biometrics.idealtest.org/dbDetailForUser.do?id=3
19. AOptix Iris Database. http://www.aoptix.com
20. The BTAS Competition on Mobile Iris Recognition. http://biometrics.idealtest.org/2016/MIR2016.jsp

Parametric Shape Descriptor Based on a Scalable Boundary-Skeleton Model

Ivan Reyer$^{(\boxtimes)}$ and Ksenia Aminova

Federal Research Center "Computer Science and Control" of RAS, Moscow, Russia
reyer@forecsys.ru, flake.inbox@gmail.com

Abstract. A parametric shape descriptor based on the relation between contour convexities and skeleton's branches of a shape is suggested. The descriptor contains the set of a polygonal figure convex vertices approximating a raster image and estimations of significance for curvature features corresponding to the vertices. The estimations are calculated with use of a boundary-skeleton shape models family generated by the polygonal figure. The applications of the suggested shape descriptor to the face profile line segmentation and content based image retrieval are described.

Keywords: Shape analysis · Boundary-skeleton shape representation · Skeleton base · Parametric shape descriptor

1 Introduction

One of the object's shape fundamental characteristics that can be used for a shape representation is an object's boundary. It was found that the human visual system analyses and identifies an object's shape using contour convexities and concavities [3,8]. Based on this principle, there is a number of structural shape analysis methods in computer vision using a contour represented as a sequence of convex and concave features [7,18]. When an object is approximated with varying accuracy, shape features manifest themselves according to their significance: the more pronounced is a feature, the more crude and general shape models will contain it. Thus, in addition to the identification of a feature it is important to estimate its significance. Therefore, a problem arises: constructing a shape descriptor containing information on features at various levels of detail.

One of the approaches to the problem is represented by corner detection methods [9,10,17]. These methods are based on the selection of local contour curvature extrema by a threshold value. In this case the significance estimation is the absolute value of a boundary fragment curvature corresponding to a feature.

Another popular tool is the curvature scale space model [1,6] based on Gaussian smoothing of a contour and location investigation of curvature extrema and zero-crossings at different levels of scale.

To estimate a raster object's boundary curvature the described approaches use either an adaptation of the curvature concept to a discrete contour representation, or the approximation of an object's boundary with a piecewise-smooth curve.

© Springer Nature Switzerland AG 2019
V. V. Strijov et al. (Eds.): IDP 2016, CCIS 794, pp. 151–162, 2019.
https://doi.org/10.1007/978-3-030-35400-8_11

In this paper, it is suggested to use the relation between contour's convex parts and branches of an object's skeleton [5] to identify and estimate curvature features. A parametric family of variously detailed boundary-skeleton shape models obtained from a polygonal figure approximating an object is considered. A boundary-skeleton model consists of the skeleton base [12,19] of the polygon and the boundary of all base disks union. An analysis of changes within the family allows us to calculate significance estimations for curvature features generated by convex vertices of the polygon's boundary. The set of convex vertices with their significance estimations is used as a shape descriptor.

2   Skeleton Base

Let's briefly describe the concept of a skeleton base presented in [12,19].

Suppose P is a simply-connected polygonal figure, and ε is a non-negative number. Let H denote the Hausdorff distance.

Definition 1. *A closed disk C is called an ε-admissible disk of P if P has a maximal inscribed disk C' such that $H(C', C' \bigcup C) \leq \varepsilon$.*

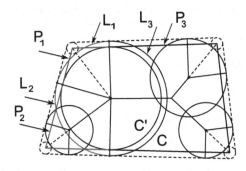

Fig. 1. Fragments of the boundary and arcs of a maximal ε-admissible disk

Definition 2. *A closed disk C is called a maximal ε-admissible disk of P if it is an ε-admissible disk of P, and there does not exist another ε-admissible disk of P $C' \neq C$ such that $C \subset C'$.*

It is quite easy to see that the set of centers of all the maximal ε-admissible disks of P is the same as the set of centers of all the maximal inscribed disks of P, i.e. the skeleton of P.

Definition 3. *Suppose C is a maximal ε-admissible disk of P, and C' is the corresponding maximal inscribed disk of P. Suppose the tangency points of C' divide the boundary of P into n fragments $P_1, P_2, \ldots, P_n$, $n \geq 2$, and the corresponding radii of C divide its boundary into n arcs $L_1, L_2, \ldots, L_n$ (Fig. 1). Then C is called a base disk of P if there are $i : 1 \leq i \leq n$, and $j : 1 \leq j \leq n$, $i \neq j$, such that $H(P_i, L_i) \geq \varepsilon$ and $H(P_j, L_j) \geq \varepsilon$.*

Definition 4. *The skeleton base of P is the set of centers of all the base disks of P.*

Therefore, the skeleton base of P is a subset of the skeleton of P.

The skeleton base of a polygonal figure defines a closed planar domain (i.e. the so-called "skeleton core" [20]) formed by a union of the base's edges vicinities (Fig. 2). These vicinities are bounded by line segments and fragments of parabolae and hyperbolae in accordance with the type of an edge. The skeleton core describes possible positions of a skeleton of any closed simple-connected region which is close to the polygon in the sense of Hausdorff distance. Therefore, the skeleton base may be considered as a stable skeletal representation of a shape.

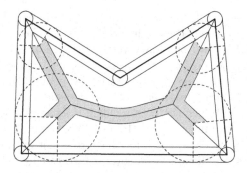

Fig. 2. Skeleton base and skeleton core of a polygonal figure

Suppose p is a skeleton point, C' is a maximal inscribed disk with center in p (see Fig. 3). C' divides the boundary into n fragments. Let's denote by U_i, $i = 1, \ldots, n$ the subsets of convex boundary vertices belonging to the fragments. Let's consider the maximum distance from p to points of U_i:

$$d_i = \max\{d(p, u)) | u \in U_i\}.$$

Suppose distances d_i, $i = 1, \ldots, n$ are arranged in ascending order:

$$d_1 \leq d_2 \leq \cdots \leq d_{n-1} \leq d_n.$$

Let's select a subset U_j with the second largest distance d_j. If there are several such sets, we are going to consider any of them. If $d_1 = d_2 = \cdots = d_{n-1} = d_n$, let's select an arbitrary U_i. Let's denote the selected subset by U', the corresponding maximum distance from p to points of U' by d', and the furthest point of U' by f.

As ε increases, a skeleton base changes monotonically and continuously [21]. The skeleton edges are erased by pairs of curves (parabolae and hyperbolae). For an edge generated by two segments of the boundary (Fig. 4a) erasing curves are two parabolae with the focus f and the directrices parallel to the segments. For a parabolic edge generated by a vertex and a segment (Fig. 4b) one erasing

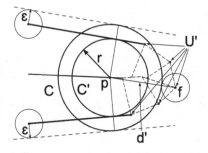

Fig. 3. Terminal vertex of a skeleton base

curve is also a parabola with the focus f and a directrix parallel to the segment, and the second one is a hyperbola with foci in f and the vertex. If an edge is a segment generated by two boundary vertices (Fig. 4c), erasing curves will be two hyperbolae with foci in f and the vertices.

It is obvious that an edge can be erased by a number of different pairs. In the first place, different fragments of an edge may have different furthest points in the same U'. Hence the points where a skeleton edge intersects edges of the furthest point Voronoi diagram [16] of U' are the change points of erasing pairs. Second, fragments of an edge may have different U' sets. In this case so called central points equidistant from several furthest points f are considered. Besides, one pair may erase an edge in two opposite directions. In this case the edge has a point where one erasing curve is tangent to the edge and the other degenerates into a ray. All these points, where either an erasing pair or a erasing direction is changed (points of intersection with furthest point Voronoi diagram, central points, tangency points), together with the original skeleton vertices, compose a skeleton markup [21]. A skeleton markup defines a marked skeleton (Fig. 5). Every edge of the marked skeleton is erased by a unique pair moving in one direction. In addition, with each edge one can associate two values of ε corresponding to the endpoints of the edge.

3   Parametric Shape Descriptor

The boundary of all base disks union can be considered as figure's contour model. This model reflects boundary features which are significant within the accuracy of approximation. As ε grows, the skeleton base loses branches generated by convex features becoming insignificant. Hence, with every convex feature one can associate a value of ε at which the corresponding branch of the skeleton is excluded from the skeleton base. This concept forms the basis for a parametric shape descriptor similar in a way to the curvature scale space models. The descriptor contains the list of all convex vertices of a polygonal figure approximating the raster object, and significance estimations for curvature features corresponding to the vertices. The significance estimations are calculated in a following way.

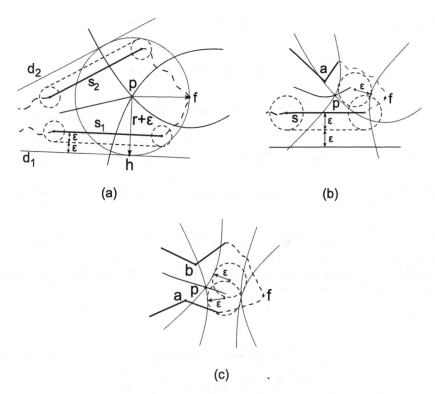

(a)   (b)

(c)

Fig. 4. Erasing curves of different edges

Let's mark the polygonal figure skeleton and consider the erasing of its branches. Here we suppose that the skeleton base does not change its connectivity as ε grows. The erasing starts at the terminal vertices of the skeleton (i.e. convex vertices of the polygon's boundary) and moves into the figure. For each edge of the marked skeleton two associated values of ε corresponding to its endpoints are defined, and the direction of erasing is known (from the point with smaller value to the bigger one). Suppose a vertex q of the marked skeleton has a degree of $n > 2$. Then it has $n-1$ incoming edges $v_1, ... v_{n-1}$ and one outcoming edge v_n with corresponding values $\varepsilon_1, ..., \varepsilon_n$. The value ε_n associated with v_n is maximal, $\varepsilon_n = \max(\varepsilon_1, ..., \varepsilon_n)$, and is equal to at least one of $\varepsilon_1, ..., \varepsilon_{n-1}$. This means that an incoming edge v' having the maximal ε will be erased last. Then, the edge v_n is associated with the branch containing v' and the erasing of the branch continues. The erasing of branches containing other incoming edges stops, and the initial convex vertices of the boundary obtain significance estimations equal to the corresponding values of ε. The process stops when the central point of the marked skeleton is reached.

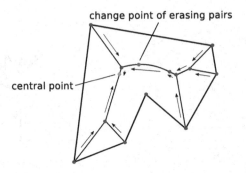

Fig. 5. Marked skeleton

For the edges q_1q_3 and q_2q_3 of the marked skeleton presented in the Fig. 6 the values of ε at q_3 are equal to $\varepsilon_1 = 23$ and $\varepsilon_2 = 1.5$, respectively. Hence q_2q_3 is erased first, and the significance of q_2 is equal to ε_2. The erasing started at q_1 continues, and the significance estimation is equal to the value of ε at the central point q_c.

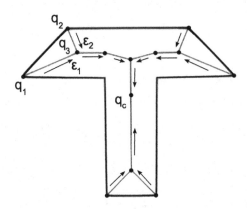

Fig. 6. Calculation of significance estimations

The Fig. 7 presents an example of a shape and its boundary convex vertices being significant for given values of accuracy.

Three types of a skeleton base's connectivity change are recognized according to the number of fragments being erased last and the kind of a final vertex: unique "central" fragment with a central point; several fragments with tangency points; unique fragment with a tangency point. The described procedure of significance estimation can be modified for these three cases. The Fig. 8 shows the parametric descriptor of a shape with 3 central points. The erasing divides the skeleton base of the polygon in points z_1 and z_2. Each of three fragments has its own central point (c_1, c_2, c_3), with c_3 being the last vertex of the skeleton base.

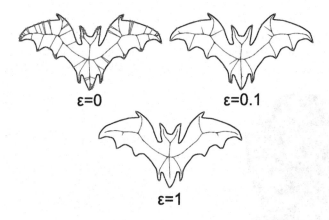

$\varepsilon=0$ $\varepsilon=0.1$

$\varepsilon=1$

Fig. 7. Significant convex vertices of a shape

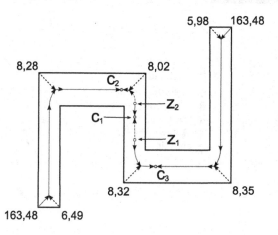

Fig. 8. Calculation of significance estimations for the figure with 3 central points

4   Experimental Results

In this section two examples of parametric descriptor usage for shape analysis and comparison are described.

The first example is the task of profile line segmentation from a face image. To reveal the profile line, a source image is transformed to a binary image of a head using brightness or color segmentation. Then the boundary-skeleton models of head and background are constructed (Fig. 9a), and the significance estima-tions for convex vertices of both models are calculated. The suggested segmen-tation scheme is based on a rather simplified model of the profile line. It does not take into account possible significant deformations of the shape caused by such details as beard or glasses.

Let's assume that the head on the image is turned to the left. Then the profile line is searched within the left part of the head contour (from the most left point with the maximal ordinate to the most left point with the minimal ordinate). Ten fiducial points denoted as $P_1, ..., P_{10}$ according to [13] are identified in this part of the contour on the basis of calculated significance estimations (Fig. 9b).

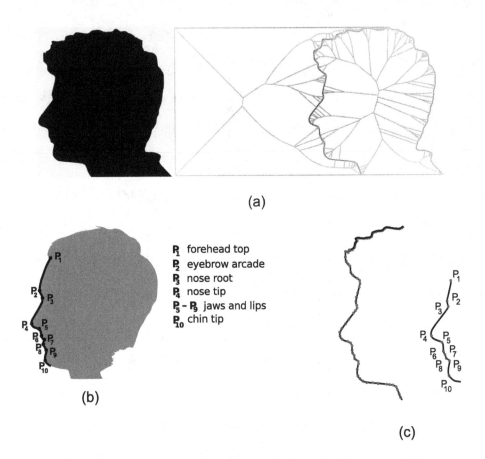

(a)

P_1 forehead top
P_2 eyebrow arcade
P_3 nose root
P_4 nose tip
P_5 - P_9 jaws and lips
P_{10} chin tip

(b)

(c)

Fig. 9. Segmentation of a profile line from a face image

The nose tip P_4 is selected from three convex vertices with the biggest significance values lying in the middle and bottom parts of the specified fragment. The vertex with the middlemost ordinate is considered as P_4. The chin tip P_{10} is searched as the convex vertex of maximal significance lying below P_4. The points $P_5, ..., P_9$ are situated between P_4 and P_{10}. The points P_6 and P_8 correspond to two convex vertices with the biggest values of significance, and the points P_5, P_7 and P_9 correspond to three concave vertices with the biggest values of significance. The nose root P_3 is searched as the concave vertex of maximal significance lying at a distance not greater than $|P_4P_{10}|$ above P_4. The eyebrow arcade P_2 is searched among the convex vertices with the biggest significance values lying at a distance not greater than $|P_3P_4|$ above P_3. The point P_1 lies within the group of concave vertices above P_2. To eliminate possible identification errors, an additional check of the order of convexities and concavities and anatomy proportions is made.

The segmentation method was tested with a combined image base consisting of 1032 faces. The base included 136 grayscale profile images of 28 people from the face database of the University of Bern [2], 716 color profile images of 372 people from the Color FERET database [14,15], and 180 color profile images of 16 people prepared by the authors (8–25 images per person with variations of the head position). Adequate sets of fiducial points were obtained for 885 images from the data set.

The identification errors generally included wrongly identified P_1, P_3, P_4 and P_2 (the errors of others points P_5–P_{10} occur 6 times, so it may be said that the suggested scheme works correctly on the bottom part of profile line). More than half of wrong results were generated because of wrongly identified P_1. The reason for such errors is that it is difficult to describe correctly the location of P_1 on the contour in terms of curvature significance. 26 errors are made in computation of P_3: most of them are the misinterpretation of the hair part points on the contour. The errors of P_4 (there are 20 of them) may occur if a head is poised forward or backward to a great extend. The reason of 15 wrongly identified P_2 is explained by the absence of eyebrow arcade. An example of the correct profile segmentation is presented in the Fig. 9c.

The second example is a comparison of color images by shape and color of segmented regions. Hand-segmented color images from Berkeley Segmentation Dataset [11] are used for the comparison. A continuous model of a segmented image consisting of nonoverlapping polygonal figures set is constructed. Each polygon from the set approximates a segmented raster region within the image, with polygons of two neighbour regions having common boundary fragments. Then, the obtained models are compared by shape and color of polygons with the use of one-to-one matching scheme [4] finding minimal average distance between polygons. To obtain equal number of polygons for compared images, adjacent polygons will be combined, if their corresponding raster regions are similar in average color. To estimate the shape similarity, the change of significant vertices' number from 10 to 2 at increase of ε is compared using L_1 distance between functions. An example of two turtle shapes comparison is presented in Fig. 10. To estimate the color similarity, CIELab color model is used. Both shape and color distances are normalized to the range [0, 50].

The comparison procedure was tested within 103 segmentations of 31 images (2–5 segmentations per image). For 83 models other segmentations of the same image were among the five most similar segmentations. In 89 cases other segmentations of the same image were among the ten most similar segmentations. Some results of the comparison are presented in Fig. 11.

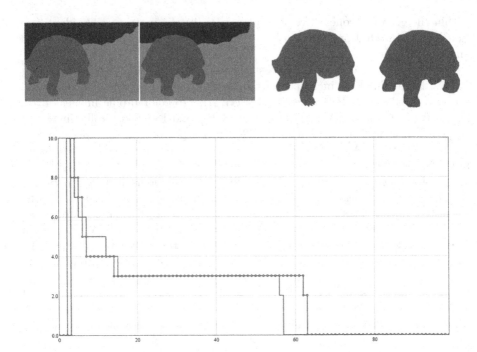

Fig. 10. Comparison of segmented regions by the change of significant vertices' number

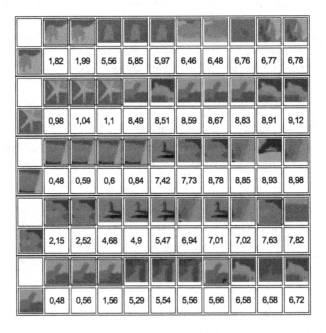

Fig. 11. Results of the segmented images comparison by regions shape and color

5 Conclusion

Monotonic and continuous change of a skeleton base with accuracy value growth allows to examine the family of skeleton bases as a whole and select skeletal representations with the appropriate accuracy of approximation. The concerned boundary-skeleton shape models family defined by skeleton bases and the parametric descriptor makes it possible to investigate curvature features at different levels of detail and identify contour fragments containing the required combination of curvature features without the approximation by piecewise smooth curves or contour smoothing. The descriptor may be used together with the marked skeleton to obtain a boundary representation of an object at the required detail level.

Acknowledgements. The research was supported by the Russian Foundation for Basic Research (projects No. 14-07-00736, 17-07-01432, 17-20-02222).

References

1. Abbasi, S., Mokhtarian, F., Kittler, J.: Curvature scale space image in shape similarity retrieval. Multimedia Syst. (1999). https://doi.org/10.1007/s005300050147
2. Achermann, B.: University of Bern face database. Copyright 1995, University of Bern, all rights reserved (1995). ftp://iamftp.unibe.ch/pub/Images/FaceImages/
3. Attneave, F.: Some informational aspects of visual perception. Psychol. Rev. **61**(3), 183–193 (1954)
4. Bartolini, I., Ciaccia, P., Patella, M.: Query processing issues in region-based image databases. Knowl. Inf. Syst. (2010). https://doi.org/10.1007/s10115-009-0257-4
5. Blum, H.: A transformation for extracting new descriptors of shape. In: Models for the Perception of Speech and Visual Form, pp. 135–143. MIT Press (1967)
6. Dudek, G., Tsotsos, J.K.: Shape representation and recognition from multiscale curvature. Comput. Vis. Image Underst. **68**(2), 170–189 (1997). https://doi.org/10.1006/cviu.1997.0533
7. Galton, A., Meathrel, R.: Qualitative outline theory. In: Proceedings of the 16th International Joint Conference on Artificial Intelligence, vol. 2, pp. 1061–1066 (1999). http://dl.acm.org/citation.cfm?id=1624312.1624370
8. Hoffman, D.D., Richards, W.A.: Parts of recognition. Cognition **18**, 65–96 (1984). https://doi.org/10.1016/0010-0277(84)90022-2
9. Koplowitz, J., Plante, S.: Corner detection for chain coded curves. Pattern Recogn. **28**(6), 843–852 (1995). https://doi.org/10.1016/0031-3203(94)00100-Z
10. Latecki, L.J., Lakamper, R.: Shape similarity measure based on correspondence of visual parts. IEEE Trans. Pattern Anal. Mach. Intell. **22**(10), 1185–1190 (2000). https://doi.org/10.1109/34.879802
11. Martin, D., Fowlkes, C., Tal, D., Malik, J.: A database of human segmented natural images and its application to evaluating segmentation algorithms and measuring ecological statistics. In: Proceedings Eighth IEEE International Conference on Computer Vision, ICCV 2001, vol. 2, pp. 416–423 (2001). https://doi.org/10.1109/ICCV.2001.937655
12. Mestetskii, L.M., Reyer, I.A.: Continuous skeletal representation of image with controllable accuracy. In: Proceedings of International Conference on Graphicon-2003, pp. 246–249 (2003, in Russian)

13. Pantic, M., Rothkrantz, L.J.M.: Facial action recognition for facial expression analysis from static face images. IEEE Trans. Syst. Man Cybern. **34**, 1449–1461 (2004). https://doi.org/10.1109/TSMCB.2004.825931
14. Phillips, P.J., Moon, H., Rizvi, S.A., Rauss, P.J.: The FERET evaluation methodology for face recognition algorithms. IEEE Trans. Pattern Anal. Mach. Intell. **22**, 1090–1104 (2000). https://doi.org/10.1109/34.879790
15. Phillips, P.J., Wechsler, H., Huang, J., Rauss, P.: The FERET database and evaluation procedure for face recognition algorithms. Image Vis. Comput. **16**(5), 295–306 (1998). https://doi.org/10.1016/S0262-8856(97)00070-X
16. Preparata, F., Shamos, M.: Computational Geometry. Springer, New York (1985)
17. Ray, B.K., Pandyan, R.: ACORD - an adaptive corner detector for planar curves. Pattern Recogn. **36**(3), 703–708 (2003). https://doi.org/10.1016/S0031-3203(02)00084-5
18. Rosin, P.L.: Multiscale representation and matching of curves using codons. CVGIP: Graph. Models Image Process. **55**(4), 286–310 (1993). https://doi.org/10.1006/cgip.1993.1020
19. Zhukova, K.V., Reyer, I.A.: Parametric family of boundary-skeletal shape models. In: Proceedings of 14-th Russian Conference on Mathematical Methods for Pattern Recognition (MMPR-14), pp. 346–350 (2009). (in Russian)
20. Zhukova, K.V., Reyer, I.A.: Structure analysis of object shape with use of skeleton core. In: Proceedings of 8-th International Conference on Intelligent Information Processing (IIP-08), pp. 350–354 (2010). (in Russian)
21. Zhukova, K.V., Reyer, I.A.: Parametric family of skeleton bases of a polygonal figure. Mach. Learn. Data Anal. **1**(4), 391–410 (2012). (in Russian)

Author Index

Printed in the United States
By Bookmasters